AF324059

Profile of Engineering Education in India
Status, Concerns and Recommendations

Profile of Engineering Education in India
Status, Concerns and Recommendations

Gautam Biswas

K.L. Chopra

C.S. Jha

D.V. Singh

Indian National Academy of Engineering

Narosa Publishing House
New Delhi   Chennai   Mumbai   Kolkata

Profile of Engineering Education in India
Status, Concerns and Recommendations
256 pgs. | 15 tbls.

Gautam Biswas
Department of Mechanical Engineering
Indian Institute of Technology, Kanpur
Presently on Deputation as Director – CMERI, Durgapur

K.L. Chopra
Former Director
Indian Institute of Technology Kharagpur

C.S. Jha[†]
Former Vice Chancellor
Banaras Hindu University, Varanasi and
Former Director, Indian Institute of Technology Kharagpur

D.V. Singh
Former Director
Indian Institute of Technology, Roorkee and
Former Vice Chancellor, University of Roorkee

NAROSA PUBLISHING HOUSE PVT. LTD.

22 Delhi Medical Association Road, Daryaganj, New Delhi 110 002
35-36 Greams Road, Thousand Lights, Chennai 600 006
306 Shiv Centre, Sector 17, Vashi, Navi Mumbai 400 703
2F-2G Shivam Chambers, 53 Syed Amir Ali Avenue, Kolkata 700 019

www.narosa.com

ISBN 978-81-8487-061-9

Published by N.K. Mehra for Narosa Publishing House Pvt. Ltd.,
22 Delhi Medical Association Road, Daryaganj, New Delhi 110 002

Printed in India

*Dedicated
to the
memory of one of the greatest educators and
engineering-scientists of contemporary India,
Prof. C.S. Jha, who is no more with us to see
the printed version of this book*

Foreword

Today there are more than 2300 Engineering Institutions in the country turning out more than 6,00,000 engineers every year. Indian Engineering Education represents one of the largest educational systems in the world. The challenges posed by this rapidly growing system in our country are extremely complex. While addressing the strategy of nurturing a large number of highly intelligent post-secondary students in a proper manner, quality of Engineering Education emerges as the issue of paramount importance.

An engineering education is relevant for the graduate if it meets the needs of the industry. A core set of such needs must be identified. It should include skills related to manufacturing, computer technology, interpersonal relations, analytical and experimental techniques. Academic programs have to be evaluated and revised periodically so they endow graduates with skills pertaining to the above-mentioned set.

Having mentioned this, the focus should be drawn on training the mind and developing the power of analysis. This can be accomplished through a proper and well balanced curriculum and appropriate grading system. It should be remembered that the primary philosophy of the teaching should be to collect, collate and feed the information to the students in a digestible form.

The engineering graduates of the advanced countries possess a high level of confidence in handling unknown problems. Many of our graduates are deficient in handling unknown problems. The gap has to be bridged through setting up appropriate educational system. The IIT model has been successful. This model should be adopted by a large number of Institutes for academics and governance. Creation of excellent educational system will eventually terminate the acrimonious debates about the reservation.

The present INAE sponsored research study is aimed at developing a comprehensive assessment of the status of engineering education in India in the light of the increased global expectations from the engineering graduates. The study targets examining challenges being faced in terms of access, equity, regional imbalance and quality. The study is also aimed at analyzing the weaknesses of the present system and identifying the requirements of modern teaching learning processes. The study suggests measures for improvement in faculty qualification and competence. Finally, the study provides close guidance in instituting healthy governance. The task force of the INAE, comprising of Prof. Gautam Biswas, Prof. K.L. Chopra, Prof. C.S. Jha and Prof. D.V. Singh, worked hard over a

period of more than a year and presented its report to the council. The council felt
that the publication of the report in the form of a book will serve a great purpose.
The document will serve as the guidance to the leaders of the Colleges, Universities, Institutions and Government to adopt the best practices.

I wish the policies pertaining to Engineering Education in the country benefit
immensely from such a sincere effort of the Academy.

Dr. P.S. Goel
FNAE, FNA, FNASc, FASc, FIETE, FAeSI, FTWAS
President
Indian National Academy of Engineering

Preface

India has emerged as a major player in the world in the field of Engineering Education, and Indian engineers have contributed significantly to the economic and technological development of many foreign lands, not only in the Information Technology sector but also in general engineering services and in hi-tech research and development in solid-state electronics, communications, and embedded systems. The present INAE sponsored study is aimed at developing a comprehensive assessment of the status of engineering education in the country in the light of the increased global expectations from engineering graduates, examining challenges being faced in terms of access, equity, regional imbalance and quality, analyzing the weaknesses of the present curricula and the requirements of modern teaching learning processes, and suggesting measures for improvement in faculty qualification and competence, system of governance and other related areas so as to bring a major transformation of the system in teaching and research.

Chapter 1 gives a historical perspective of the origin and growth of engineering related education in the country with very significant and often forgotten detail from the times of the ancient Buddhist Viharas through the British period and up to modern times both before and after independence. After a thorough and insightful journey through the history, the chapter concludes the need of most equitable, sustainable and effective system for modern and developed India.

Chapter 2 enumerates the changing global scene on the engineering education front in terms of alternative delivery mechanisms, new accreditation criteria, emerging social concerns, impact of globalization and emergence of the global engineer. Details of two major studies carried out by the US National Academy of Engineering in terms of Attributes of Engineers of 2020 and Educating the engineers of 2020 are brought in to give the current international perceptions of the changing needs of the future. The recommendations of the Global Engineering Excellence study are also mentioned to strengthen the futuristic outlook. It is felt that Engineering Education is continuously evolving in all parts of the world to meet the new challenges of the emerging knowledge society and the gradual integration of the world economies. Its emphasis on quality assurance through output/outcome based criteria and through developing attributes which would enable graduates to work in a global environment would permeate to all countries and help restructuring engineering curricula towards increased innovation and creativity and equipping all graduates with co-operative and harmonious working in multidisciplinary and multicultural teams.

Chapter 3 examines the challenges currently facing Indian engineering education. The rapid expansion of institutions and intake capacity is welcome but may not meet the government's intention of raising the Gross Enrolment ratio of Higher Education from the present 10-11% to 15% by the end of the 11th plan and 21% by the end of the 12th plan as this would in itself need a higher rate of growth for engineering education. International comparison of Science & Technology personnel per 1000 of the population is shown to put India in very poor light even when compared with Asian countries (India 3.5, China 8.1, South Korea 45.9, Israel 76, and Japan 110). Issue of equity involving students from poor economic background, from rural areas and from disadvantaged groups like SC/ST/OBC and gender parity are seen to be areas of concern both for the central and state governments. Regional imbalance of engineering educational facilities is an area of even bigger concern. The facilities for engineering education are so badly distributed in the country, that 63.6% of the UG seats and 62% of the PG Seats are available in only three of the seven regions of the country, namely the Southern, the South Western and the Western. Further, nearly 60% of engineering graduates and post-graduates come from only 4 states of Tamilnadu, Andhra Pradesh, Karnataka and Maharashtra. Facilities for engineering degree education varies in the major states from a low of 1.86 per one hundred thousand in Bihar to a high of 189.20 per hundred thousand in Tamilnadu, and for post graduate engineering education from a low of 0.2 per hundred thousand in Bihar to a high of 10.94 per hundred thousand in Tamilnadu. Some strategic state intervention has been recommended to reduce this imbalance in the next 10 years. The recent large-scale expansion in engineering education has come at the cost of the quality of the educational offerings. The affiliating University system, outdated curricula, inadequate academic infrastructure, shortage of qualified teachers, poor teaching/learning processes, lack of innovative and creative activities, archaic evaluation system, absence of proper academic ambience, and non-participative and bureaucratic governance mechanisms have all contributed in different degrees to the lowering of standard of the educational offering of a large proportion of the 1700 or so engineering institutions in the country. Many suggestions are given to improve the quality assurance mechanism and setting up quality bench marks including up gradation of the competence of teachers and the learning ambience in institutions.

Chapter 4 deals with the sensitive issue of ethics of engineering practice A professional is expected to contribute to the progress and well-being of society in a sustainable environment through creation and dissemination of knowledge, and by using engineering knowledge and skills to develop technologies, products and services, as also regulatory mechanisms, etc consistent with an accepted ethical code of conduct. The evolving tech-nomic-globalisation demands adherence to internationally acceptable ethical values and code of conduct. It is, therefore, essential for academic institutions to develop the mindset of engineering students to cultivate a sense of social responsibility to uphold the honour and dignity of the profession in a global environment. The profession that is aimed at service and achieving excellence demands unconditional adherence to defined codes of practice and to its moral and ethical values.

Chapter 5 addresses the issue of faculty shortage across the country and the reasons for failure to attract and retain a large number of talented professionals in teaching career. Some strategies are recommended for upgradation of qualifications of existing faculty including a revision of existing norms regarding number of seats and the amount of scholarships of QIP fellowships, appointment of adjunct faculty for sharing expertise with industry and some important policy changes to be adopted by the Government. A need is expressed for institutionalizing faculty development mechanism. The most effective way to address the problem of faculty shortage is considered to be the expansion of postgraduate education and research many fold.

Chapter 6 examines the important issue of curriculum reforms. Since engineering education is expected to produce trained manpower for maintaining and advancing technological growth in the country, the relevance and usefulness of engineering education have a strong bearing on the country's technological capability. The educational strategy should assist in transforming the Indian economy to a knowledge economy. The systems involved in this endeavor should strive for furtherance of knowledge. The Universities and Institutes have a major responsibility of analyzing and understanding the state of existing knowledge and its future perspectives, creating new knowledge, and assimilating the new knowledge that is being generated world over. The students have to understand the effect of globalization on the society. The Academic Programmes should have enough strength and flexibility to cross the disciplinary boundaries, and to play with innovative concepts. A well balanced curriculum with the suggestive structure of common curriculum for all departments of an Institute (such as basic sciences, mathematics, language courses, humanities, basic engineering, and engineering practices viz., graphics, workshops, etc.) has been outlined. The model curriculum also strikes a pedagogic balance with the department specific compulsory components, electives and open electives (with the option of Minor).

Chapter 7 deals with the post graduate and research programs in Engineering The objectives of a healthy postgraduate programme in engineering are the development of a more general and fundamental understanding involving mathematical and natural sciences; more general and more powerful methods of analysis; capacity to understand the advanced work, through which the field is advancing; and nurturing imagination, and intellectual capacity to make new advancements.

Chapter 8 addresses teaching learning processes and their improvement. Cognitive learning is demonstrated by knowledge recall and the intellectual skills: comprehending information, organizing ideas, analyzing and synthesizing data, applying knowledge, choosing among alternatives in problem-solving and evaluating ideas or actions. This domain on the acquisition and use of knowledge has to be predominant in the majority of courses. The evaluation mechanism should be well designed to test all the aspects of cognitive learning. Every Institute/ University should be able to make a critical assessment of its own strengths and weaknesses over a period of three years. There should be sincere efforts from the Institute/ University to correct the weakness.

Chapter 9 examines in detail the question of governance of engineering institutions. The growing concerns for value for money, efficiency, excellence, greater access, quality assurance, competitiveness, accountability, etc have altered the way the education is to be managed. Changes in strategies to control ways of monitoring, assuring and assessing the quality of education and accountability are universal requirements. The governance and management of technical institutions in India is ill-defined, inflexible, highly bureaucratic, demotivating, unaccountable, and caste and politics driven. The need to reform our higher educational system drastically has been emphasized by several reports to the government. A model governance framework must provide academic, administrative, and financial autonomy with built-in measures for checks and balances and accountability to usher in a culture based on *"institute of academics, for academics, by academics"*.

Research and Development and Institute-Industry interaction are discussed in Chapter 10. The economic growth of any society is now increasingly dependent on creativity through human resources, innovation through research and development, and capital through intellectual property rights. Consequently, besides diffusion and assimilation of knowledge, creation and utilization of knowledge through research and development must be an integral component of any engineering education. A very large majority of our engineering institutions have failed to evolve from undergraduate teaching to post graduate teaching and research. The vast majority of private engineering institutions are primarily first degree level ones and has made little effort to introduce research oriented programmes for economic reasons as also for lack of qualified faculty. The engineering education system in India has been unable to attract the best engineering students for post-graduate studies. Even the IITs are unable to attract their own engineering graduates for R&D and have failed to meet the objectives of research oriented institutions. A grave national challenge for academic institutions, government Science & Technology funding agencies and industries is to interact and collaborate together in a new era of liberal, flexible and pragmatic governance and financial rules and regulations for conducting relevant R&D with meaningful outcome. Collaboration between industry and academia needs to be viewed as an asset and a future investment by all associated stake holders.

Chapter 11 describes Technology Assisted Learning systems. It is shown that proper use of Information and Communication Technologies (ICT) can make quite a few problems in education simple. In the teaching-learning environments in the educational institutions, it is necessary to enhance the effectiveness of learning. The ICT should be deployed in a big way by the teachers and students. The courses created under the auspices of NPTEL (National Programme on Technology Enhanced Learning) should be used to supplement learning. With advancements in the communications and data management techniques, it is possible to disseminate and share information across the country. These advances need to be utilized to improve the efficiency of the system.

Innovation, intellectual property, and entrepreneurship are closely linked to the engineering profession, since engineers improve existing engineering devices, services and systems (innovation), build completely new businesses, artifacts and systems (intellectual property), and set up new businesses in construction, manufacturing or services (entrepreneurship). These issues are examined in some detail in Chapter 12. The engineering education system can play a major role in equipping graduates with skills of innovation and creativity, training them in identifying and protecting their intellectual property rights, and making them familiar with the requirements, characteristics and methodology for successful entrepreneurship. Growth of global economies and life styles of civilized societies are being increasingly determined by knowledge and innovation created and nurtured by knowledge institutions. The demand for a closer interaction of such institutions with entrepreneurs, communities and industry is increasingly becoming louder. Entrepreneurship strengthens the knowledge system, converts knowledge to intellectual property, promotes techno-ventures for commercialization of technologies, creates wealth and enhances technology competitiveness and the tech-image of the country. Further, it creates new business and creative opportunities, jobs and services, and thus promotes regional as well as local development. Close interaction of academic institutions with entrepreneurs should be mandated as a societal responsibility on the part of academia.

Chapter 13 describes in detail the new initiatives being taken both by Governments and the private sector as well institutions to meet the challenges facing engineering education in the country and assist in developing a viable, vibrant and quality conscious engineering education system. The recommendations of the National Knowledge Commission, the Eleventh Five-year Plan proposals, the educational initiatives of Infosys, Wipro, and TCS, the emerging importance of Indo-US collaboration in Engineering Education (IUCEE), and the distance education schemes of IITs are highlighted among others so as show that the whole nation is gearing up to meet the growing Indian economy's requirement of high quality trained manpower.

The policies on education, the actions and interventions of the regulatory authorities under the provisions of the Acts and the gazetted Regulations and the judicial verdicts affect the dispensation of the education in the country and its quality and standard at all its levels. In Chapter 14, some policy issues, various provisions of Acts and Regulations and some judicial verdicts have been discussed limiting to higher engineering and technical education to emphasize that a more comprehensive and in-depth analysis is required to re-visit the National Education Policy and to realize the necessity to evolve an altogether new policy, which addresses itself to the contemporary national and international scenario, emerging WTO-GATS regime, mobility of students, scholars and teachers and other issues for imparting quality education.

In Chapter 15, the conclusions are drawn about the entire study and the consolidated recommendations are suggested.

Gautam Biswas
K.L. Chopra
C.S. Jha
D.V. Singh

Acknowledgements

The INAE sponsored research study on *"Profile of Engineering Education in India: Status, Concerns and Recommendations"* was undertaken by us in early 2008. We conducted our study for more than a year and had several deliberations. The present book is the outcome of our joint efforts. We express our debt of gratitude to all the members of the Governing Council of INAE in general and to the President and the Vice Presidents in particular for taking enormous interest in the task we had undertaken. They also participated in several rounds of discussions with us during the course of preparation of the manuscript. The Secretariat of INAE has always extended a helping hand for the smooth conduct of the research study.

We express our gratitude to Prof. Rangan Banerjee and Prof. Vinayak P. Muley of the Indian Institute of Technology Bombay for helping us with the information as and when it was needed. As such, the review report of Prof. Banerjee and Prof. Muley, entitled "Engineering Education in India" became a pre-cursor of our study. We are deeply indebted to Prof. Prem Krishna, one of the Vice Presidents of INAE, who reviewed the entire script of the book and quietly inspired us with his brilliance. The suggestions received from him were very useful. The cheerful voice of Prof. R. Natarajan, former Vice President of the INAE, was a constant source of encouragement. The insight received during the discussions with the Fellows at the INAE Seminar organized by Prof. Natarajan held at Bangalore, in February 2009 was very helpful. We record our appreciation for the critical comments received from Dr. Baldev Raj of IGCAR on the preliminary version of our research report. One of us (GB) was greatly benefited from the discussions with Prof. G. Neelakantan of IIT Kanpur and Prof. Sreerup Raychaudhuri of TIFR. It is ubiquitous that we are indebted to all the authors of the references that were studied by us during execution of the research.

Gautam Biswas
K.L. Chopra
C.S. Jha
D.V. Singh

Contents

Foreword   *vii*

Preface   *ix*

Acknowledgements   *xv*

1. **Historical Glimpses of Technical Education in India**   1
   D.V. Singh

2. **Engineering Education with International Perspective**   22
   C.S. Jha

3. **Challenges Faced by Indian Engineering Education**   41
   C.S. Jha

4. **Ethics in Engineering Education**   61
   K.L. Chopra

5. **Faculty Related Issues**   82
   Gautam Biswas

6. **Undergraduate Curriculum and Evaluation Mechanisms**   90
   Gautam Biswas

7. **Postgraduate Education and Curriculum**   101
   Gautam Biswas

8. **Improving Teaching-Learning Processes**   107
   Gautam Biswas

9. **Governance and Management of Technical Institutions**   113
   K.L. Chopra

10. **Promotion of Research and Industrial Interaction**   126
    K.L. Chopra

11. **Technology Assisted Learning**   139
     Gautam Biswas

12. **Innovation, Intellectual Property Rights, and**   144
     Entrepreneurship
     C.S. Jha and K.L. Chopra

13. **National Initiatives**   161
     C.S. Jha

14. **Technical Education — Policies, Acts, Regulations**   193
     and Verdicts
     D.V. Singh

15. **Consolidated Recommendations**   220

Annexures   228

About the Authors   235

1

Historical Glimpses of Technical Education in India

"We should remember that immediately on attainment of freedom our people are not going to secure happiness. As we become independent all the defects in the system of elections, injustice, the tyranny of the richer classes and also the burden of running administration are bound to come upon us. People will begin to feel that during those (British) days there was more justice, there was better administration, there was peace, and there was honesty to a very great extent among the administrators compared to the days after independence. But there is hope, if education spreads throughout the country and people develop from their childhood qualities of pure conduct, God fearing and love. Otherwise India would become an abode of grave injustice and tyranny of the Rulers."

— Mahatma Gandhi

1.1 EARLY CENTURIES

In India the Buddhists developed University institutions several hundred years before similar institutions of higher learning (Studium Generale) appeared in the 12th Century in Europe. The Buddhist Viharas in India developed into bigger institutions such as Nalanda, Vallabhi, Vikramshila, Odantapuri, Jagaddala, Mithila, Ranchi, etc. However, these Universities disappeared due to decline of Buddhism in India.

The excellence of manufactured articles in medieval India, e.g., fabrics of cotton and silk, embroideries, painted and enameled wares, steel guns, swords, knives and scissors, gold and silver ornaments and white paper, is well known. This excellence was achieved and maintained for centuries with dependable technical education practices comprising hereditary learning, pupilage training and training schools attached to workshops. The manufacturing establishments called *Karkhanas*, imparted technical education in their areas of specialization. The early Sultans and Mughals supported such Karkhanas and their technical/vocational education. Both

Hindus and Muslims took great interest in vocational education as a result of which trained workers of every trade were available in abundance.

"Government will not fail to employ education, to strengthen its hands, and perpetuate its institutions."
                                                                    — William Godwin

1.2   TECHNICAL EDUCATION IN BRITISH INDIA

In 1794, the British Government in India established the first Survey School in Madras with eight students from English schools. The Madras Survey School then trained only English boys. The British policy at that time was against teaching surveying to native Indians because of military and political implications of survey work, as a precaution against reliable maps falling into the hands of the French, the Dutch and the Portugese. The Court of Directors of the East India Company insisted on the secrecy of survey maps and restricted the art and science of surveying to English boys. However, civil surveying for revenue purposes remained outside the ambit of the restrictions of the East India Company because it was an ancient branch of knowledge in India and a class of people, the 'Amins' or 'Mirdhas', specialized in preparing land revenue maps. The Madras Survey School went through several ups and downs, was on the brink of closure in 1810 but was revived in 1819, admitting some apprentices directly called from England in addition to boys from the local English schools. The Survey School was later expanded in 1857 and renamed as the Civil Engineering School.

1.3   DEVELOPMENTS IN BENGAL

The General Committee of Public Instruction, comprising mostly English officers, constituted in Bengal in 1823, remained for about twenty years the only agency of Bengal Government concerned with education, until it was replaced by a Council of Education in 1842. The Committee and its successor Council in their reports frequently mentioned the branches of study which would be useful to students to earn their livelihood. Apart from reading, writing, and arithmetic, surveying was recommended for Indians required in judicial and revenue departments and by courts. At that time surveying was taught in Bengal in two colleges, the Mohammedan College (established in 1781) and the Hindu College (established in 1817).

On the occasion of the establishment of the Sanskrit College in Calcutta, Raja Ram Mohan Roy wrote to Lord Amherst in 1823 and urged the Government to *"employ European gentlemen of talent and education to interest the natives of India in Mathematics, Natural Philosophy, Chemistry, Anatomy, and other useful sciences, which the natives of Europe have carried to a degree of perfection that has raised them above the inhabitants of the other parts of the world."* The opinion that crystallized in Bengal was that drawing and surveying should be taught only in colleges and not in schools. From the need to teach these subjects in colleges, the desirability of having colleges of civil engineering was a big step. Engineering was not classified into several subdivisions and it meant engineering for civil purposes as distinct from military. The importance of civil engineering, as a branch of instruction for Indians, began to attract attention of authorities in about 1843. Construction of roads and canals was in progress or being projected and it was realized that men trained in engineering would be required and surveying skills alone would not be enough. At this time the idea for having a University was gaining ground. The Council in its report in 1844-45 suggested the establishment of a central university for "granting degrees in arts, science, law, medicine and civil engineering". However, the Bengal Government took no action until after ten years.

1.4 DEVELOPMENTS IN BOMBAY AND NWP

Elphinstone Institution in Bombay started engineering classes in 1844 to train surveyors and builders. The programme had a short life of just three years. It closed because it did not attract students of sufficient merit to qualify as "scientific civil engineers fully groomed in the theory of their art".

In early 19th century, after the conquest of the northern region by the British, the North Western Province (NWP) was created in 1836 with Agra as its headquarters. In 1843 the subject of education in NWP was transferred from the Bengal Government to the Government in Agra. The renovation work on old Yamuna Canals and construction of some roads were undertaken. Thinking had started on the Ganga Canal. Appointment of James Thomason as the Lieutenant Governor of NWP at that time proved to be very significant. While the Governors of Presidencies, appointed from the British aristocracy and political heavy weights, responded slowly through proper channels to the proposals coming from lower levels, Thomason rising from the ranks of civil servants, made the proposal himself, followed them up for acceptance by higher authorities and implemented them vigorously.

Thomason submitted a detailed proposal for setting up a college to train Indians as Civil Engineers. Probey Cautley, an army engineer whose name is closely associated with the construction of Ganga canal, had envisaged a school for the supply of efficient workmen for the entire line of canal. Thomason went further in his proposal dwelling extensively on the requirements of the whole country regarding surveying, irrigation, navigation, roads, bridges and railways for all of which it was impossible to provide Europeans. He made a strong case to form a nucleus at Roorkee for training Civil Engineers, which led to establishment of the Civil Engineering College at Roorkee in 1847 with Lieutenant Maclagan as the Principal and four teachers, two of whom were Indians. The College began with three courses, one for engineers, one for upper subordinates (overseers) and the third for lower subordinates (sub-overseers and draughtsman). The engineering class was open to European military officers and English, Anglo-Indians and Indian civilians. During the first twenty years, only a few Indians were admitted. But after 1870 the number of Indians increased and that of Europeans declined.

The Roorkee College became a model and catalyst for the other engineering colleges in India. Within a few years, engineering colleges of Calcutta, Madras and Poona followed. Thomason died untimely in 1853. The Roorkee College was named Thomason College of Civil Engineering in 1854 in honour of its founder. The concept of the engineering education through formal instructions in a school or college was unknown at that time even in England. Therefore, the subjects of engineering, when Roorkee College was established, were not properly classified from the point of view of teaching. The teachers of Roorkee College, within the first twenty five years of its existence did pioneering work in systemizing the teaching of engineering, formally writing lecture notes, developing examples, making drawings, writing books and manuals and updating them periodically. The Roorkee College Manuals and Treatise on Civil Engineering became standard texts not only in Roorkee but at other colleges in India.

After a rather stagnant period of Roorkee College during the decade of 1890, programmes expanded to include mechanical and electrical engineers at one end to motor drivers and linemen at the other. Machinery and tools were indented from England and a modern workshop was established, including forge and foundry shops run entirely on electricity. Several new courses were added, including a course in telegraphy in 1893. In 1897, the two year engineering course was extended to three years with two branches; civil and electrical, telegraphy was merged with electrical engineering. The mechanical apprentice and the industrial apprentice courses of three years were opened one year earlier to train foremen. The industrial apprentice course covered various trades, printing, photography, metal and woodwork, electric motors and mechanical engines. During the early decades of the 20th Century, the profile of the college changed from degree-level education to an industrial apprentice level institute. Even an automobile driver's class was started in 1907.

1.5 WOOD'S DISPATCH

The famous Wood's Dispatch of July 1854 from the Court of Directors of the East India Company envisaged an enlarged system of education to be pursued in India in view of the need of suitably trained persons for PWD in all the three Presidencies. In response, the Governor General Lord Dalhousie recommended to the Court for establishment of an engineering class at each of the three Presidencies of Calcutta, Madras and Bombay. Unlike that in NWP, the progress was slow but things started moving.

In Madras by 1842 it was already recognized that the survey school was inadequate for the needs of PWD and establishment of an engineering college was desirable. The initial reaction of the Court of Directors to the proposal was negative on the ground that the general education in the Presidency had not advanced enough to justify such a venture. It took the Court five years to agree with the Supreme Government on the need to upgrade the Survey School. The Director of Public Instruction (DPI) Madras recommended the establishment of a college on the pattern of Roorkee College.

In 1854, Lord Dalhousie proposed that on the principle of Thomason College at Roorkee, a complete system of instruction should be provided at Madras for every class belonging to Europeans of PWD and natives, whether artificers, foremen, overseers, surveyors or civil engineers. The Court of Directors conveyed their concurrence in 1855. The Director of Public Instructions (DPI) submitted a detailed scheme of an institution at three levels just as at Roorkee College, with provision of military students along with civilians. The Madras Government approved the scheme with the provision that every student of the college be required to master some craft or trade and the school at the Gun Carriage Factory in Madras be extended to supply the PWD with artificers. The Supreme Government eventually accepted the proposal of Madras Government with some stipulations. The Civil Engineering College came into being in Madras in 1859 but functioned on a very modest scale because Madras Government gave no grants in addition to what was given to the Survey School. The College was affiliated to the University of Madras and the first batch received the Bachelor of Civil Engineering (BCE) in 1864. The Madras University added mechanical engineering course in 1894 and the name of the degree was changed from BCE to BE. The Madras College was the first to offer a degree programme in mechanical engineering. In 1920 the college moved to Guindy.

1.6 FLUID STATE IN BENGAL

In 1854, the Council of Education Bengal recommended the establishment of a separate engineering department in the proposed Presidency College, which was to function in the following year by incorporating the Hindu College. The Chief Engineer of Bengal, Col. Goodwyn, however, recommended the constitution of a separate engineering college for general improvement of the Department of Public Works. The Bengal Government concurred with the Chief Engineer, but took two years to prepare its detailed scheme to obtain the sanction of the Court of Directors. In November 1856, the Civil Engineering College Calcutta started functioning with 10 students and two teachers. The Government gave a guarantee of employment in PWD to the students after passing the two year course. The College was affiliated to Calcutta University when it was established in 1857 and the course was raised to three years with one year training at the end of which the candidates got the degree of Licentiate in Civil Engineering (LCE).

In 1865 the Engineering College was merged as a department of Civil Engineering with the Presidency College, which impeded the progress of engineering education in Bengal for the next about fifteen years. After a gloomy existence, the prospects improved in 1878 when a practical training institute around a newly established PWD Workshop was established and it was decided that the civil and mechanical engineering students should receive theoretical training in a college and practical training in a workshop. In April 1880 the engineering department in the Presidency College was again given a separate identity as a college and was moved to Sibpur. It was given the name Government Engineering College Howrah. Later in May 1887 it got another name Civil Engineering College Sibpur. In 1920 it was named as Bengal Engineering College Sibpur and remained so until after independence.

1.7 THE POONA COLLEGE

On the proposal of the Bombay Government, the Supreme Government in 1855 approved the establishment of a college for the instruction of civil engineers, surveyors, overseers, foremen and artisans. The project started in a low key with an Engineering Class and a Mechanical School at Poona. Later Sir Cowasjee Jehangir made a magnificent donation of Rs. 50,000 which helped in converting the School into Poona Civil Engineering College in 1864 and affiliation to University of Bombay in 1868 for the degree of Licentiate in Civil Engineering. The scope of instruction of the college was expanded in 1879 to include a forestry course for forest rangers and a diploma course in agriculture. The name of the College of Engineering was changed to the College of Science. General courses in science were started leading to B.Sc. degree of Bombay University. The most distinguished and famous alumnus of the College, Sir Mokshagundam Visvesvaraya topped the list of L.C.E's in 1884.

In 1886, the course of civil engineering was revised and upgraded and the minimum qualification for admission was raised from matriculation to what was then called Previous Examination. The trend towards opening new courses was reversed. The Rangers Class was closed in 1904. By 1911 all other non-engineering courses were also abolished. The B.Sc. classes were transferred to the Science Institute in Bombay and the College was given back its original name, the Poona College of Engineering.

1.8 MILITARY HEGEMONY OVER THE ENGINEERING COLLEGES

The Royal Engineers in the army in British India played a major role in influencing the fortunes and also the misfortunes of the four engineering colleges. They were the only type of engineers that came to India with the East India Company. As the Company took more and more responsibility of governance of the country, all technical jobs of engineering and scientific nature were entrusted to the military engineers. PWD was almost entirely officered by them. Engineering education fell naturally in their sphere. The original proposal for each of the four engineering colleges was to place them under the education departments of the respective Presidency or Province to fulfill the civilian objectives of the colleges to train engineering personnel for civilian work. But many of the Principals, Professors and Instructors appointed to organize and run these colleges were military engineers, who forged close links between the college and the military in two ways. The army became, at least partly, the feeder for admission in the college and it served the needs of the army by running tailor made special courses for officers and other ranks of the engineering corps.

From the beginning, the Roorkee College was placed under PWD and not under the education department as envisaged originally. The Government of India recognized that the Thomason College had special obligations towards the army and that a certain proportion of seats in the engineering and the upper subordinate classes should be reserved for young men from the army who wanted to change over to services in public works. Also, under the orders of the Military Department of the Government of India, special courses were regularly run for army personnel till the end of the 19th Century. All the Principals, except one, were military officers until 1930.

The other three colleges successfully remained under the control of the department of education. However, in matters of admission and courses, Madras College had similarity with Roorkee College. Its last Royal Engineer Principal retired in 1907. The Calcutta College too showed military bias but its life was short. The engineering classes of the Presidency College and later Sibpur College had no military connection. The first Principal of Poona College Capt. JBH Close was very keen to make the College serve military needs by providing instruction to European sappers posted in Bombay Presidency. The Education Department did not allow too much importance to training military officers at Poona. Only the first two Principals were from the military. Subsequently, the military connection ceased.

1.9 STANLEY ENGINEERS AND COOPER'S HILL COLLEGE

By 1858, it was clear that the ambitious objectives of the Roorkee College envisaged by its founder did not quite fit in with the imperial interests perceived by Her Majesty's Government in England, which had taken the responsibility of directly governing India after the 1857 war of independence. A scheme of directly recruiting young Civil Engineers in England for service in the public works in India was initiated by the Secretary of State Lord Stanley in 1859. Engineers recruited under the scheme were dubbed as Stanley Engineers. From 1859 to 1868, 165 'Stanley Engineers' were appointed against only 45 from Roorkee College.

While the military engineers nurtured all the four engineering colleges in their infancy, within a few years they also initiated policies which adversely affected the growth of the colleges. In 1870, as a modification of the Stanley Scheme, a college of civil engineering, Royal Indian Engineering College, popularly called Cooper's Hill College, on the pattern of Roorkee College, was established in England to train engineers for PWD in India. The entire expenditure of Cooper's Hill College was borne by the Government of India and its purpose overlapped with those of the engineering colleges in India. Persons recruited and trained in England thus filled bulk of the posts in India. The products of the engineering colleges in India were deprived of job opportunities, with the result that many went unemployed or were under-employed. The persons behind the Stanley Scheme and the Cooper's Hill College were some military engineers who were earlier associated with the establishment and running of the four colleges in India. A major share of appointments in PWD in India went to candidates from Cooper's Hill. Later due to fall in demand for engineers in PWD, the recruitment to PWD from Cooper's Hill was reduced in 1892 and thereafter. The Cooper's Hill College limped on for another 17 years and was finally closed in 1907.

1.10   ESTABLISHMENT OF UNIVERSITIES

In July 1854, the Court of Directors of the East India Company sent a Dispatch to the Governor General of India in the Council suggesting the establishment of Universities in the three Presidencies. Consequently, the Universities of Calcutta, Madras and Bombay were founded in 1857. At that time, they were purely examining bodies concerned mainly with instituting degrees and framing rules and regulations for examinations leading to degrees. Their scope covered all branches of learning involving intellectual efforts worthy of recognition by a University. These Universities established faculties and instituted degrees in traditional areas like arts, science, law, etc. and also in medicine and engineering.

It was left to colleges to enroll students, organize teaching and seek affiliation with the university degrees. For engineering, Calcutta and Bombay Universities aimed higher and instituted Master of Civil Engineering (M.C.E) degree for which a B.A. degree was prerequisite. Madras University more realistically instituted, in addition to M.C.E., a lower degree called Graduate in Civil Engineering (GCE) for which prerequisite qualification was matriculation. The three engineering colleges at Calcutta, Madras and Poona were duly affiliated to the Universities of their Presidencies.

The needs of engineering education and the academic aspects through which the colleges were required to steer, did not match the expectations of the Universities. The Colleges continued to produce engineering personnel at two levels, officers and subordinates with great efficiency. If the Universities did not consider the engineer officer worthy of university degrees, the college certificates were valuable enough. There were no takers for the MCE degree. Therefore, Calcutta and Bombay Universalities like Madras University, instituted a lower degree of L.C.E. From 1864, the graduates of engineering colleges at Calcutta, Poona and Madras received university degrees, LCE and BCE. The University of Allahabad was established in 1887 and the Punjab University in 1892. The Roorkee College was affiliated to Calcutta University in 1864 and to Allahabad University in 1894. These affiliations remained only notional as there is no record of Roorkee students taking university examination. The affiliation with Allahabad University ended in 1905.

1.11 STATUS OF TECHNICAL EDUCATION (1884-85)

Sir MacDonnell's Memorandum prepared in 1886 on the existing state and future prospects of technical education had nothing much to notice except the four engineering colleges, three schools of industrial art and about forty five lower grade industrial schools. Sir MacDonnell reported that a few more Survey Schools were functioning during the period in Hyderabad (Sind). The total enrollment in the four engineering colleges during 1884-85 was reported to be 608, that in Survey Schools 465 and in industrial schools, 1379.

> *"The aim of public education is not to spread enlightenment at all: it is simply to reduce as many individuals as possible to the same safe level, to breed a standard citizenry, to put down dissent and originality."*
> — H.L. Mencken

1.12 GOLDEN JUBILEE OF QUEEN'S REIGN

The golden jubilee celebration of Queen Victoria's reign in 1887 provided the occasion to collect large sums of money from the public for commemorating the event suitably. In Bombay and Madras opening of technical schools was considered an appropriate way of utilizing the funds. Viceroy Dufferin in a public address in Lucknow in 1888 said *"I have always taken the deepest interest in technical education. I have called the attention of all the subordinate governments to the desirability of promoting this branch of instruction by every means in their power. But I must remind you that it is neither within the competence nor the function of the Supreme Government to give practical effect to its views. The latter responsibility devolves on the local government to a certain degree, but still more largely on the Indian community. Even the local governments, unassisted by the liberality and counsels of those who are in a position to support their efforts, can but do little"*.

The Victoria Jubilee Technical Institute, founded in 1887, proudly claims that the impulses, which led to the formation of Indian National Congress, were also responsible for the origin of the Institute. The great names associated with the Institute include Dadabhai Naoroji, Pherozshah Mehta, Dinsha Wacha, R.D. Ranade, Badruddin Tyabji, K.T. Telang, and Philanthropic members of Jeejabhai, Wadia and Petit families. The public donations collected to commemorate the viceroyalty and to celebrate the Golden Jubilee of Victoria's reign provided the funds.

1.13   AT THE END OF 19TH CENTURY

The facilities of technical education at the end of 19th Century consisted of 4 engineering colleges at degree level, about 20 survey and technical institutions and some 50 industrial schools. The standard of education was good except that in the industrial schools. The diplomas/ degrees of the colleges were recognized by the Institution of Civil Engineers, London. However, there was stagnation in the growth of engineering education and a decline in the number of students passing out of the four engineering colleges during the last almost four decades of the 19th Century. The reason for non-utilization of the capacity was not the non-availability of suitable candidates but the then prevailing employment policy of the Government. Government support for the development of technical education had come to a standstill by 1875. In the next quarter of the century, the Engineering Colleges barely continued to exist. The attitude of the Government was at variance with the need of the public. An elaborate system of technical education was being demanded to improve the employability of youth and to reduce the poverty of the people, whereas the Government opinion was that the development of industry was a pre-requisite to the expansion of technical education. The resulting debate helped in creating public awareness of private efforts. VJTI was the first example. The national sentiment for technical education found its greatest expression in the endowment of Rs.30 Lakh created by Jamshedji Nusserwanji Tata in 1898 to promote technical education, which finally led to the creation of IISc in Bangalore.

1.14   EMERGENCE OF PUBLIC OPINION

In 1880s, a new class of educated Indians was growing who had exposure to education, literature and political thoughts of the West. This class wanted to emulate the West and progress in respect of technical education and industrial development. This new class began to impress on the Government the need to start and accelerate the pace of such development. It did have the support of some Englishmen but the Englishmen on the official side said that the Indians were trying to move too fast. The newspapers with national leanings were continually faulting the

Government for not making enough provisions for technical education. Unemployment was on the increase and technical education was viewed not only as the basic need for industrialization but also as a means to ameliorate unemployment.

The Indian National Congress in its third Session held at Madras in 1887 passed a resolution that "— *having regard for poverty of the people, it is desirable that the Government be moved to elaborate a system of technical education* ". A mention of the need to have technical education became a regular feature of the Presidential address at Congress sessions and its resolutions. In particular, the subject was emphasized by the Congress Presidents, W.C. Banerjee in 1892, Ananda Mohan Bose in 1898, Chandavarkar in 1900 and Madan Mohan Malviya in 1909. Earlier in the 1901 session the Congress passed a resolution expressing its opinion that a Government College of Mining Engineering be established at some suitable place on the model of the Royal School of Mines in England and the mining colleges of Japan and Europe.

Individuals were coming forward with schemes of their own and published pamphlets and newspaper articles emphasizing the need of technical education. At the same time some individual Englishmen and English newspapers derided these ideas. For example, one of them bemoaned that *Bombay had already voted money for technical education to loosen a further flood of technical education upon this land; the bulk of native educated community approved of the scheme because they did not understand what it meant.* Even the prestigious Pioneer, then a leading Anglo-Indian paper observed, *"technical education in Central Provinces being everywhere rampant, and that engineering and agricultural classes have sprung up with the rapidity of mushrooms"*

"Men had better be without education than be educated by their rulers."
— Thomas Hodgskin

1.15   LORD CURZON'S DISPENSATION

In 1900 Viceroy Lord Curzon appointed Sir Edward Buck to advise him on technical and industrial schools. Buck recommended that technical education be separated from the general education at all levels in terms of separate schools and separate administrative set-ups. His views however, were ignored. In 1901, Curzon appointed a committee under the chairmanship of Col John Clibborn, Principal of Thomason College to examine the subject of technical education. Its report did not appeal to Curzon and was not even published. Curzon did not favour spreading higher level technical education in India. His response to Tata's scheme of starting a higher level science research institute was lukewarm. Curzon was of the opinion that India did not have the necessary educational base to profit from higher technical education. He was in favour of beginning at the lowest level.

Fresh Government resolutions resulted only in opening a few more technical and industrial schools. There were few industries in India and most of them were owned by Europeans. They preferred to employ Europeans to all technical positions. In 1902, Indian Universities Commission was appointed, which made its recommendations on technical education: *"As the Colleges of Engineering train a large number of students for the lower branches of the profession and only a small number for the higher branch, of which alone, the University takes cognizance, we do not think it desirable that the University should itself undertake instruction in Engineering"*

1.16 PROVINCES INVITED TO MAKE SUGGESTIONS

The Government of India sent out a circular in 1903 to provincial Governments asking for suggestions on utilization of an additional grant for technical education. Punjab replied that it wanted no immediate establishment of technical schools. United Provinces, as advised by the Upper India Chamber of Commerce, considered technical schools unnecessary and that the industrial schools served no useful purpose. Madras wanted no grants for expansion. Bombay wanted to reorganize the VJTI and to start three or four new technical schools. Bengal wanted a weaving school at Serampore. These negative replies further convinced Curzon that technical education was not needed.

The Indian public opinion was, however, different. When Curzon fund exceeding Rs.1.5 Crores for setting up a Memorial for the Queen Victoria was collected, the public opinion was strongly in favour of utilizing it to advance the cause of technical education. It was suggested that the fund be added to Tata's gift for a science institute. But Curzon preferred a monument in stone. The Swadeshi Movement of the first decade of the 20th Century led to the urge of Swadeshi education also. During this period, many national educational institutions, free from the Government control, were established as for example, National Council of Education, Bengal (1906), which later became Jadavpur University and some others. A few of- them also imparted technical education. In the decades that followed, some more institutions imparting technical and industrial education at various levels and in different disciplines of technology were established.

1.17 POST WORLD-WAR 1

After the World War I, the Government became a little more responsive to public demand of technical education. The situation, which was stagnant during the War period, began to improve. Several institutions were established during the War and in the decades thereafter. They include Banaras (then Benares) Hindu University (1916), Harcourt Butler Technology Institute, Kanpur (1920), Calcutta University College of Science and Technology (1920), Bihar Engineering College, Patna (1924), Indian School of Mines, Dhanbad (1926), Maclagan College of

Engineering, Lahore (1930), Andhra University, Vizag (1933), University Department of Chemical Technology, Bombay (1934), and Aligargh Muslim University (1935). Some Colleges were started in the Princely States of India in 1937. Many other colleges in the four regions of the country were started in subsequent years.

There was a lack of coordination at the all India level and to some extent at the provincial level on the issues of contents and durations of the educational programmes. The nomenclatures, "Engineering', 'Technical', 'Technological', or 'School', 'College', and 'Institute' were arbitrarily used and did not indicate the level of the programmes. This has not changed even to-day.

Abbott and Wood report commissioned by the Government of India in 1937 stated that provision for technical education at all levels was too inadequate for a large country like India and that there was no integrated policy and there was no coordinating agency for its proper development. Later for coordination and standardization of courses, the All India Association of Principals of Technical Institution was formed in 1941. Several technical institutions were established in different parts of the country during this period. As recommended by the Abbot-Wood Committee, a Polytechnic was established in Delhi in 1941.

1.18 PRE-INDEPENDENCE INITIATIVES

The number of engineering colleges during the years before the independence was 46 with a total intake capacity of 2500 students. These colleges catered predominantly to the needs of the various Government departments such as Public Works, Railways, Electricity, Telecommunications, Irrigation, etc. A very small proportion of engineers found opportunities in private sector companies engaged in engineering operations and productions.

In 1944, the Central Advisory Board of Education was asked by the Reconstruction Committee of Viceroy's Executive Council to give a Report on the post-war education development in India. In the light of the Report, the Council appointed a committee in 1945 under the chairmanship of N.R. Sarkar to consider the development of higher technical institutions in India. In its interim report submitted in 1945, the Sarkar Committee recommended the establishment of not less than four Higher Technical Institutions one each in the North, East, South and the West. The objectives of these institutions were expected to be similar to those pursued by the Massachusetts Institute of Technology in the USA. The key features of the proposed programme were non-specialized orientation and integrated curricula supported by institutional processes that would encourage Indian students to think creatively. The products of these institutions were expected to be "creative scientist-engineers" and technical leaders with a broad human outlook and individuals with "creative initiative in future situations".

All students were expected to have strong core knowledge of basic sciences, engineering sciences, humanities, and technical arts besides the professional courses in their chosen disciplines.

On the recommendations of the Sarkar Committee, a national agency, All India Council for Technical Education (AICTE, then not a Statutory body), was established in 1945 for planned and coordinated growth of technical education in India. In 1947 at the time of Independence, Polytechnic education was lacking severely in respect of eligibility, duration, standard and management. At that time, only 53 institutions conducted Diploma courses in the country with an intake capacity of 3670 students. The major task of AICTE was therefore, coordination, standardization and improvement of Polytechnic education.

The action on the other recommendations of the Sarkar Committee, which led to establishment of IITs, was taken by the Government of India after Independence. The visionary report of the Sarkar Committee led to the birth of the first Indian Institute of Technology at Kharagpur in 1951 followed by four other IITs at Bombay, Madras, Delhi and Kanpur in the late fifties and early sixties. The Abid Hussain Committee, which was constituted to review the IITs, clearly recommended that more IITs be not established and the intake of the existing IITs be increased to meet the demand of quality technical institutions. In view of Rajiv Gandhi's Assam Accord and due to political considerations, the IIT at Guwahati, Assam was established which started functioning in 1992. Narasimha Rao explained that the new IIT was an IIT with a difference. On the historic occasion of reorganization of three States in 2001, when the new State of Uttaranchal was born, circumstances became favourable to convert University of Roorkee into an IIT, an event which was soon made an election plank by many.

"Public educators, like Soviet farmers, lack any incentive to produce results, innovate, to be efficient, to make the kinds of difficult changes that private firms operating in a competitive market must make to survive."   — Carolyn Lochhead

1.19   EARLIER INITIATIVES IN INDEPENDENT INDIA

The Government of India, after Independence, recognized the importance of quality technical education for the economic and industrial growth of the country, which required the future generation of engineers to be competent, innovative, good designers and excellent product manufacturers. The Radhakrishnan Commission Report (1949) made several recommendations on technical education emphasizing the need of new types of engineering and technical institutions in India. As a consequence of these recommendations, several new developments took place in the subsequent years.

The initiatives to conceive a grand design for technical education in India and implement the concept true to its spirit, in the form of Indian Institutes of technology (IITs) is one of the greatest hallmarks of visionary development in independent India. The performance of the IITs during the last four to five decades has provided a sense of fulfillment. Their achievements, in producing high quality engineers and technologists and in undertaking advanced R&D missions, have received worldwide recognition. Though their future directions may depart from the original design, there is every hope that they will continue to maintain their reputation for excellence and their brand image.

Considering the ambitious economic and social development goals of post independent India, the need for training engineers in much larger numbers and in diverse disciplines was acutely felt. While a number of engineering Colleges were started adopting the conventional pattern of engineering education there was a desire to establish a few technological institutions which would provide World class technical education and would have the dynamic characteristics to adopt themselves to rapid changes in engineering knowledge and its applications.

In 1959, under the chairmanship of M.S. Thacker, a Working Group on Technical and Vocational Training and also a Committee on Postgraduate Education and Research were constituted. Another important step in the development of technical education was the appointment of Kothari Commission (1964). The Kothari Commission Report contained many recommendations, which included the importance of practical training and industry-institute interaction.

During the first two decades after Independence, the expansion in technical education was accompanied by massive investments by the Government in infrastructure in the academic institutions. In the late fifties, faced with a serious shortage of faculty in technical institutions, a Technical teachers Training (TTT) programme was started. Reasonably attractive fellowships were offered to fresh graduates to induce them to study for a post-graduate degree in one of the established institutions in the country and later join the teaching profession. Some teacher trainees were also sent abroad for acquiring higher qualifications.

"All who have meditated on the art of governing mankind have been convinced that the fate of empires depends on the education of youth."   — Aristotle

1.20 GROWTH OF TECHNICAL EDUCATION

On the recommendation of the Engineering Personnel Committee, which was appointed by the Planning Commission in 1955, the Government of India initially decided to establish eight Regional Engineering Colleges (RECs). In order to provide each State with a Regional Engineering College, seven more were approved for establishment during the Third Plan period, bringing the total number of

RECs to fifteen by 1972. Two more were added later, one in Jullandhar (1985) and the other in Hamirpur (1989).

The Technical Teacher Training (TTT) programme, which was started in the late Fifties, resulted in the creation of a large pool of dedicated teachers. After the TTT programme was phased out, the Quality Improvement Programme (QIP) for improving the quality of technical education and developing the faculty of engineering institutions was launched by the Government of India in 1970. This programme is operated in 25 engineering/technical institutions; seven of them are major QIP centres and eighteen are minor centres. This programme provides opportunity for continuous upgrading of knowledge and skills of persons who are already in the teaching profession. Over the years a large number of teachers acquired higher degrees from leading institutions in the country under the programme.

In order to assess the impact of foreign technical assistance on the development of technical education in India and to determine the areas needing to be further developed and supported through the foreign technical assistance programme, the Government of India appointed a Committee under the chairmanship of Y. Nayudamma (1978). Another Committee also under the chairmanship of Nayudamma was constituted at about the same time to review the postgraduate education research in engineering and technology and make recommendations for further development. Whereas, little is known about the outcome of the former, several recommendations of the latter were implemented.

The National Policy on Education (NPE 1986) was a major development in the field of education. NPE came out at a time when the role and impact of private institutions imparting technical education were not known or even perceived. NPE was, therefore silent on this aspect and so was the resulting Programme of Action (POA 1992).

1.21 POSTGRADUATE EDUCATION IN ENGINEERING AND TECHNOLOGY

The intake capacity for Post Graduate education in engineering was a mere 30 students in 1947. Most had to go abroad to obtain postgraduate education in engineering. The postgraduate education in engineering and technology in independent India had a late start. It started in a few institutions in early 1950's but the doctoral programmes were not common until early 1960's. The recommendations of the Thacker Committee (1959-61) and of the Nayudamma Committee (1978-79) played a role in the development of the postgraduate education. These Committees were constituted by the then Ministry of Education and their reports were submitted directly to the Ministry for perusal and action.

Later a Review Committee on Postgraduate Education in Engineering was constituted by AICTE in 1995 under the chairmanship of P. Rama Rao, which submitted its Report to AICTE in 1999. The actions on this Report have been slow and sporadic. The postgraduate education in India remains weak and needs urgent attention. With a weak postgraduate education in engineering, the technology base will be weak and India will not be able to become a front runner in the field of technology, industrial productivity and the service sector, which determine the growth and development of the country.

1.22   RAMA RAO COMMITTEE REPORT

Rama Rao Committee supported the 'GATE' system of admission, recommended the increase of the duration of M.E./M.Tech programme from 18 months to 21 months, and enhancing the scholarship to postgraduate students, with a provision of its periodic review. The duration of the programme was, however increased to 24 months. The Rama Rao Committee recommended strengthening the one year postgraduate diploma programmes in suitable disciplines with industrial/ application orientation (maintenance of thermal power stations, plastics engineering, industrial engineering, VLSI design, CAD, Information Technology, et al). These programmes could be aimed at key industry personnel and the concerned industries should be expected to invest in setting up links with selected institutions. The Diploma programme could be offered both on-campus and in Distance Education mode. The Rama Rao Committee recommended enrolling foreign students to post graduate programmes, particularly in engineering areas since many countries may not have the facilities in those areas. The Committee emphasized an assured placement through active linkages with potential employers. It recommended that programmes in new areas be started after careful considerations of all aspects and with due care and listed 35 representative areas. The Committee recommended restructuring and at the same time even phasing out of the outdated programmes.

1.23 EXPANSION OF TECHNICAL EDUCATION AND THE PRIVATE SECTOR

The technical education in India has expanded enormously. A major portion of this expansion is in the private sector, with an annual intake of more than 700,000 in AICTE approved institutions. The quality of education however, is a major concern. More IITs were planned to provide quality technical education. Joshi Committee was constituted to identify the institutions which could be converted into IIT, since converting an existing institution into IIT would be much less expensive than establishing new IITs. Joshi Committee had two ex-Directors of IIT and the Committee apparently did not want proliferation of IITs. It short-listed seven institutions but recommended that they be raised to the level of IITs and further recommended that another Committee be constituted to visit the seven institutions and prepare a road map for raising them to a higher level. Consequently, the Anandakrishnan Committee did a detailed exercise which included visits to the seven institutions and submitted its report to MHRD in February 2006. The Committee stressed that these institutions should constitute a new group with a design and profile which is in the context of the world milieu of 2006 and should be distinct from IITs, which were established more than five decades ago. The Committee made bold and far reaching recommendations to enable the new group of Institutions to attain very high standards. For want of a better name the Anandakrishnan Committee suggested that they be named 'Indian Institute of Engineering Science and Technology (IIEST), a name which implied greater emphasis on science. The response of MHRD on the implementation of the Anandakrishnan Committee report was initially ad-hoc in terms of a lump sum grant to the seven institutions towards the end of the financial year and later very slow.

In the mean time MHRD announced the establishment of eight new IITs and conversion of IT BHU into IIT. Six of the new IITs were started in 2008 under the mentorship of the existing IITs, IIT Punjab being mentored by IIT Delhi; IIT Rajasthan by IIT Kanpur, IIT Madhya Pradesh by IIT Bombay, IIT Orissa by IIT Kharagpur, IIT Patna by IIT Guwahati and IIT Andhra Pradesh by IIT Madras. In 2009, IIT Himachal is to start functioning under the mentorship of IIT Roorkee and IIT Gujrat under IIT Bombay. IT BHU is likely to be converted into IIT in 2009. To begin with the Director and the Chairman of the Board of Governors of each mentoring IIT were respectively appointed as the Director and the Chairman of its mentored IIT. Recently the Directors have been appointed for the new IITs which are functioning in make shift/temporary premises. Teaching is largely being done by the faculty of mentoring IITs who commute from one campus to other. Time has come for policy makers to look beyond IITs on the lines as was recommended by the Anandakrishnan Committee under the earlier reference, especially since the new institutions being established in 2008 and later in 2009 with huge investments, need not be copies of the institutions which were established more than fifty years ago even though they are the mentors.

"Scientists investigate that which already is; Engineers create that which has never been."
— Albert Einstein

1.24   INDIA WAS A LATE STARTER

N R Dhar once remarked that "The Dark Age in Medieval Europe lasted for 300 years, whilst in India it persisted for 1200 years from the year 700 to 1800 A.D. when Jagdish Chandra Bose, Prafulla Chandra Ray started experimental science." Bose in late 19th Century did pioneering research, perfectly developing millimeter wave transmission and Galena detector for reception and microwaves. Ray did monumental work in Chemistry. He was a pioneering researcher and entrepreneur providing guidance and inspiration to establish chemical industries. These scientific land mark events were marvels of technology. Such events or achievements are very few and far between in India. There are almost no technological inventions or innovations which have captured the World market to give a first in the market advantage to the country. This can change only if the concept, core and structure of the technology education in India are transformed.

"Our progress as a nation can be no swifter than our progress in education. The human mind is our fundamental resource."   — John F. Kennedy

1.25 CONCLUSIONS

The development of a nation basically depends on the availability of properly trained S&T human resources that man industries and projects, produce value-added products, provide quality services, carry out R&D and create wealth. The economical survival of the country will depend on the competence of human resources and the competitive edge provided by their ingenuity. This will be possible only through effective policies and programmes for human resources development and their management (HRD and HRM). Quality education therefore, deserves very high priority.

The products of higher education that do not meet the high standards of quality represent poor return for the investment made by the society in education and cause degeneration of the society. On the other hand, a quality system, in which all sections of society can not participate, will be difficult to sustain. The quality of education can not be achieved and maintained without ensuring the quality of teachers and quality of students in terms of intellectual abilities, aptitudes and level of preparation. To ensure this, modalities and mechanisms would have to be evolved which are acceptable to the society and are also workable to achieve the seemingly conflicting goals of quality and social justice. The present scheme of reservations is passive and leaves ill prepared students to struggle through the education system, first in the schools and if they survive, then through the higher education in which they may sink or swim. A scheme, therefore, must be evolved which will in a declared period, make it possible that only capable and well prepared students are admitted to higher education and that this would not result in any section of society being left out. If this does not happen then the quality of education will remain an unachievable goal. To achieve this goal, reforms must be brought in the primary and the school education on a very high priority. Teachers who would man the system in future are the product of the same system today.

The only remedy is to encourage the Universities and other institutions of higher learning to acquire distinctive character with their distinctive strengths and ethos. A great majority of teachers will excel in their areas of interest if the system has the capacity and freedom to recognize merit and reward it.

One of the dilemmas of our education system which defies solution is how to reform the system which is based on restrictive teaching methods and curricula that, notwithstanding occasional changes, remain outdated and outmoded. Rapid industrial growth and its sustenance require significant improvement in education standards, particularly science and technology education. If the economy is to grow and expand, it must also have the capacity to force reform of the education system which has continued to keep the academic institutions under undue bureaucratic control stifling new ideas which could transform the education system. The institutions in the country have widely varying structure, some have a greater degree of autonomy and some others, due to their affiliation to Universities, do not have the needed functional autonomy. The quality of teaching and research can only grow in a milieu of academic freedom. Scholarship is essential to give teaching its lively edge and enlighten it but it must be recognized that teaching puts narrow research into a broad perspective and makes it meaningful. The two are essential to each other.

> *"A University stands for Humanism, for an Adventure of Ideas and for Search for Truth. It stands for the onward March of the Human Race towards even higher objectives. If the Universities discharge their duties adequately, then all will be well with the Nation and the people."*   — Jawaharlal Nehru

Having had an opportunity for more than sixty years since Independence to shape and reshape the University system, India does not seem to have reached anywhere near the goal of Universities discharging their duties adequately as visioned by Nehru. It is never too late to retrospect if all is well with the nation and if not so, whether one of the reasons is that the Universities are not able to reach the goal of discharging their duties adequately? The goal actually can or need never be achieved. What is basically important is that the Universities should keep moving towards that goal and keep reorienting their directions in consonance with the time and the changing needs of the nation.

REFERENCES

1. MacDonnell, A.P., Memorandum of Technical Education in India Prior to 1886, July 1886.

2. Indian Education Policy, 1904 — Resolution Issued by the Governor General in Council, March 1904, Calcutta 1904.

3. Review of Education in India, uinquennial Review, 1882-1937.

4. Report of the University Education Commission, 1948, Ministry of Education, Government of India.

5.	College of Engineering Pune, Centenary Souvenir, 1954; Published by the College of Engineering, Pune.

6.	History of Higher Education in South India, Vol. II, 1857-1957; Published by University of Madras.

7.	Bengal Engineering College, 125th Anniversary Souvenir, 1981; Published by Bengal Engineering College.

8.	National Education Policy 1986, MHRD, Government of India.

9.	Mittal K.V., History of Thomosan College of Engineering 1847-1949; Published by University of Roorkee.

10.	Indian Institute of Science Bangalore, Golden Jubilee Souvenir 1909-1959 and Platinum Jubilee Souvenir 1909-1984, Published by IISc.

11.	Technical Education in Independent India 1947-1997; Published by AICTE, 1999.

12.	A Walk through History, 50 Years of IIT Kharagpur; Published by IIT Kharagpur, 2002.

13.	Anandakrishnan Committee Report (submitted to MHRD), 2006.

2

Engineering Education with International Perspective

"A nation, like a person, has a mind — a mind that must be kept alert, that must know itself, that understands the hopes and needs of its neighbors — all the other nations that live within the narrowing circle of the World."

— Franklin D. Roosevelt

2.1 INTRODUCTION

Engineering Education is currently at crossroads. It can continue to function through face-to-face interaction at University type institutions and campuses or adopt a mixture of alternative knowledge delivery mechanisms using modern information technology tools and permit both synchronous and asynchronous learning through distance- and web-based education mechanisms. It can remain highly discipline/specialization oriented or become more and more a multi- and cross-disciplinary learning endeavour permitting students to recognize the fact that real life professional engineering activity involves tackling multi-disciplinary problems. It can concentrate on developing only technological excellence among its graduates or equipping them also with professional and business skills of quality and competitiveness in the design, fabrication and utilization of engineering products, services and systems. Globalization of economic activities, and mobility of professionals across national borders and the emergence of a knowledge society are putting new insights on the competency requirements of engineers, their role in the emerging international scenario, and their ability to work harmoniously and effectively in multi-disciplinary and multi-cultural teams.

Engineering graduates today require not only adequate technological ability and problem solving skills, but must also be endowed with soft skills like cooperative working, communication and presentation skills, business ethics, and inter-personal relationships, and possess a deep commitment to safety, reliability, quality and sustainability of all engineering activities in which they take part. Engineering Institutions have now a new responsibility to provide opportunities to every student to acquire these abilities in addition to their technical knowledge.

The 20th Century witnessed a phenomenal rate of growth and advance in technologies. There are no signs that the rate of growth in the present century would be any less. The half-life of many technologies are already much lower than the time span of conventional 4-year first-degree program. A fixed 4-year curriculum is already becoming obsolete, as changes have to be made frequently during the 4-year period to keep the students abreast of the latest advances in their field of activity. The rapidity of technological obsolescence is compelling the educational system to ensure that students during their stay in the Institutions develop an attitude for life-long learning and acquire self-learning skills and are provided enough opportunities to demonstrate their ability in information search and retrieval and for acquiring new knowledge. Relationship with their alma mater need no longer be a one-shot affair for students but should continue throughout their entire professional life through mentoring from their teachers, participation in continuing education programmes and access to advanced education and training courses.

The role of the teacher is changing rapidly with the realization that active learning takes place only when the students themselves contribute to the learning process with the teacher acting more like a facilitator helping students in absorbing new ideas and like a moderator in harmonizing different points of view expressed by the student learners. This change of emphasis from teaching to learning is already creating new problems of identifying what is to be learnt in the classroom and what could be left to the self-learning mode or to be acquired through laboratory and project work, assignments and the like.

2.2 SOCIAL CONCERNS OF ENGINEERING

Engineering is the profession in which knowledge of mathematical and natural sciences, gained by study, experience and practice is applied with judgment to develop ways to utilize economically the materials and forces of nature for the benefit of mankind. Engineers turn ideas into reality; they create useful products and systems through playing with imagination and possibilities, leading to new and meaningful connections and outcomes while interacting with ideas, people and the environment. Since in their professional activities engineers are constantly interacting both with people/society, the environment and their awareness of societal concerns alone can help them in becoming successful in tackling and solving technological problems faced by society. It is interesting to note that many of the social concerns that we in India face are not very different from what the IEEE fellows identified as needing substantial technological resources in a global survey conducted in 2003: energy development, fight against terrorism, environmental protection, waste management, transportation, technological literacy, digital divide, and protection of intellectual property rights. Although the IEEE list is tilted heavily towards the US perspective, if one adds inequitable distribution of benefits of economic growth among the population and the growing difference in living standards between the North and the South, and inequality in trade practices among nations, one gets a fair global picture of current societal concerns which in some way or the other is going to affect all the future engineering and technological activities. Apart from the global concerns, India must give the highest priority to the removal of poverty and illiteracy from the country and to empower the common man to take full advantage of the development process. It is really unfortunate that in spite of more than sixty years of independence from the colonial past, India still has some 30% of its people below the poverty line and one third of all the illiterates of the world live in India. The task before the engineering fraternity in building a new India free from hunger and illiteracy is indeed very daunting but the faith of the society in the ability of its scientists, technologists and economists will urge them to solve these and other problems. For this is needed the commitment, co-operation and dedication of the people to the task of reconstruction of our country and our economy even at some personal sacrifice.

2.2.1 New Accreditation Criteria and Industry requirement of Graduate Competence

Most countries in the World are changing the requirements for accreditation of degree level engineering programs in the light of the changing requirements of engineering graduate abilities. As a supplier of quality engineering manpower to the World, Indian Universities and Institutions must be aware of the output/ outcome criteria for accreditation now prevailing in most countries and ensure that graduating students do not suffer in comparison because equivalence of qualifications can become a stick to beat them with in global competition for services. The American Accreditation Board for Engineering and Technology (ABET2000) now demands that all engineering programs must demonstrate that their graduates have:

(a) An ability to apply knowledge of mathematics, science, and engineering.
(b) An ability to design and conduct experiments, as well as to analyze and interpret data.
(c) An ability to design a system, component, or process to meet desired need.
(d) An ability to function on multi-disciplinary teams.
(e) An ability to identify, formulate, and solve engineering problems.

(f)   An understanding of professional and ethical responsibility.

(g)   An ability to communicate effectively.

(h)   The broad education necessary to understand the impact of engineering solutions in a global and societal context.

(i)   A recognition of the need for, and an ability to engage in life-long learning.

(j)   A knowledge of contemporary industry.

(k)   An ability to use the techniques, skills, and modern engineering tools necessary engineering practice.

ABET also insists that every program must have an evaluation process with documented results and that action must be taken to use the results for further improvement of the program The British Engineering Professors' Council have developed their new Graduate Output Standard which demands that engineering graduates should demonstrate their

- Ability to exercise key skills in the completion of engineering tasks: communication, IT, working with others, problem solving, improving own learning and performance.
- Ability to transform existing systems to conceptual models.
- Ability to transform conceptual models into determinable models.
- Ability to use determinable models to obtain system specifications.
- Ability to select optimum specifications and create physical model.
- Ability to apply the results from physical models to create real target systems.
- Ability to critically review real target systems and personal performance.

While the accreditation agencies lay down what they consider the requirements of engineering degree that Institutions should prepare for, the employers perception for the qualities they would like to find in their graduate employees are not much different. Nokia has classified essential and desirable attributes separately as follows:

Essential Attributes

- A good grasp of basic principles of core technical material.
- An ability to apply analytical thinking to determine effective solution.
- A practical appreciation of the application of theory derived from (i) sound project work, (ii) industrial placement experience.
- Understanding and appreciation of the effect of real components and operating conditions.
- Familiarity with measurement procedures and basic test and measurement equipment.
- The capability to plan and organize one's own work efficiently.
- Good written and oral communication skills.

Desirable Attributes

- Familiarity with the design process.
- Appreciation of statistical methods.
- Commercial awareness.
- Additional technical skills, such as high level programming.
- Involvement in extra-curricular activities.
- A positive disposition/outlook/attitude/adaptability.

Nokia has specified that they do not expect that their graduate entrants would be fully rounded experienced engineers or ready made managers, or would possess high competence in specific skills or tools or are highly gifted academically and have full appreciation of volume manufacturing. Most employer's value sound knowledge of science and engineering fundamentals, capacity for teamwork, communication and learning skills, understanding of the world around them and innovative attitude. Many of these are soft skills which are rarely emphasized in University curricula to the detriment of the employability of their graduates.

2.2.2 Impact of Globalization

During the first half of the 20th Century there was a famous dictum that while science is universal, engineering is nation-specific because engineers are required to solve local problems with local, social and economic constraints. With globalization and the gradual integration of the Indian economy with the global economy, this definition is no longer valid. India should produce engineers who are capable of offering engineering expertise and services to any part of the globe, who would work in international teams with team members having multidisciplinary expertise and diverse cultural backgrounds and who would be capable of recognizing diversity of cultural preferences, differing social and economic constraints, and different engineering standards of the client nation and/ or client engineering organization. While membership of WTO and agreement under GATS entitles free mobility of Indian professionals in all member nations of WTO, it also allows professionals from other countries to offer their professional services in India. This enables healthy competition among professionals and demands that the engineering and management education system should have the quality of its output comparable to the very best in the world. This has in turn brought great responsibility on all our Universities and Colleges to ensure that their graduates possess knowledge, skills, abilities and attitudes, which will give them an edge over professionals of other countries. Universities and Institutions of Engineering and Technology have to strive hard to provide this quality to their students through modern and up-to-date curricula, hiring of competent and trained faculty, upgrading the qualifications of existing faculty through part-time and fulltime post-graduate programs and introducing the best teaching/ learning practices. Indian engineers who now have presence in all parts of the Globe could contribute to these efforts by pointing out deficiencies in the system and the changing requirements of the specialized profession as seen by them in their professional work in different countries.

"Globalization is described as 'the flow of technology, economy, knowledge, people, values and ideas across borders. Globalization affects each country in a different way due to each nation's individual history, traditions, cultures, resources and priorities' Globalization is a multifaceted process with economic, social, political and cultural implications for higher education. It poses new challenges at a time when nation-states are no longer the sole providers of higher education and the academic community no longer holds the monopoly on decision-making in education. Such challenges not only address issues of access, equity, funding and quality but also those of national sovereignty, cultural diversity, poverty and sustainable development. A further and even more fundamental concern is that the emergence of cross border higher education provision and trade in education services bring education within the realm of the market and that this may seriously affect the capacity of the state to regulate higher education within a public policy perspective. Declining policy capacity of the State could affect weaker and poorer nations and benefit the more prosperous ones. It is impossible to discuss the impact of globalization on higher education without referring to the internationalization of higher education. These two terms are often mistakenly used interchangeably. Globalization is a phenomenon which is having an impact on higher education and internationalization is interpreted as one of the ways in which higher education is responding to the opportunities and challenges of globalization. Internationalization includes a broad range of elements such as curriculum, teaching/learning, research, institutional agreements, student/faculty mobility, development cooperation and many more." (**UNESCO** Education Position Paper on "Higher Education in a Globalized Society", UNESCO 2004).

2.2.3 Changing Scenario of Engineering Education

2.2.3.1 *Merging of Disciplines*

Most of engineering activities today are multi-or trans- disciplinary and engineering education must encourage students to combine their domain knowledge with sound appreciation and familiarity with related disciplines and with the ability to recognize problems of integration of their domain activities with those of engineers of other disciplines. New advances in nano-electronics, mems, molecular computing, bio-informatics, and genetic engineering, etc. have been possible through integration of knowledge gained in quantum mechanics, particle physics, material science, solid state electronics and biological systems, etc. Without sufficient knowledge of other related disciplines, effective communication and collaboration would become increasingly difficult in the multidisciplinary teams which will develop the next generation of products and services. Emergence of Information and Communications Technology (ICT) as an integrating tool for domain specific engineering education has been a blessing to engineering educators and this tool should be more effectively utilized by curricula developers in engineering education. Students working in multidisciplinary teams for their final year project work could help in appreciating the multi-displinarity of all engineering activities and enthuse other students in cooperative working with students of other disciplines.

2.2.3.2 *New Learning Paradigm and Alternative Delivery Systems*

The Indian education system has been accustomed to teacher-centric learning where the teacher is assumed to be all knowledgeable to transmit information and knowledge via his lectures to the students in general. Some interactivity is added through question answer interventions but it is strictly limited both in time and in the coverage of the number of learners. Tutorials and Laboratory work do bring some more interactivity between the teacher and the learner but learning continues to be teacher–propelled. This model of learning is currently under serious questioning. The alternative of student-centric learning is being actively pursued in many institutions across the globe where the teacher acts primarily as a facilitator in the learning process and students become active participants in the classroom. This new model encourages self-learning and trains students to seek and validate information independently and then work co-operatively with other fellow students to absorb new knowledge and skills under the guidance, help and assistance of the teacher. Problem-based learning, project-based learning, case study method, using laboratory as a place for solving unknown engineering problems by experi-mentation, and design simulations are other tools being progressively used for imbibing students with desirable attributes and developing in them both problem solving skills and attitude for life-long learning.

2.2.3.3 *Concern for Environment and Sustainability*

Twentieth Century technologies have often created environmental degradation, acid rain, global warming, ozone layer depletion and a host of other ecological problems, stretching our natural resources to precarious levels. Impression has grown that economic development and environmental preservation are often mutually exclusive and incompatible. This notion has to be removed from the minds of young graduates through training in sustainable technologies and benign manufacturing processes which will support a healthy economy and a healthy environment. Engineers in all their project formulation are legally required today to do an environmental impact analysis before the Project can be initiated. Proper training in such impact analysis must be part of all engineering curricula.

2.2.3.4 *Alternative Delivery Modes*

Rapid advances in information and communications technology have helped the fostering of several alternative modes of education delivery and a large number of education providers have appeared on the scene. While this gives an enormous choice to the learner, the efficacy of the various alternative modes is different and suit differently to new learners and professionals. Face-to-face, distance, on-line and web-based courses and competency based certification are alternatives whose number, value, efficacy and efficiency of providing quality education or skill training are advancing day-by day. While India is still largely committed to face-to-face teaching, distance and web-based courses are becoming increasingly available to Indian students. There is a need for strict quality control of these courses as quite often they are substandard. Indian Institutions must come together in developing course wares for undergraduate and postgraduate programs for marketing not only domestically but also in the international arena as these modes of education and skill delivery are fast becoming the major source of educational delivery systems and a trillion dollar industry.

2.3 ENGINEERS OF THE FUTURE

Globalization is radically changing the way national economies around the world design, produce, distribute, and consume goods and services. Engineers are in the midst of this dynamic development. They need to work in teams and on projects with members from different nations and continents. They need to be internationally mobile, whether physically or virtually to respond to the growing international concern about the preparation of engineering workforce in the future. In a major study carried out by the U.S. National Academy of Engineering (NAE) on "The Engineers of 2020: Visions of Engineering in the New century" (Published by the National Academic Press), a detailed analysis was done to identify the types of activities which will be in the domain of the engineers in the future. According to this study "The engineer of 2020 will be faced with myriad challenges, creating aggresive and defensive solutions at the macro- and micro scales in preparation for possible dramatic changes in the world. Engineers will be expected to anticipate and prepare for potential catastrophes such as biological terrorism; water and food contamination; infrastructure damage to roads, bridges, buildings, and the electricity grid; and communications breakdown in the internet, telephony, radio, and television. Engineers will be asked to create solutions that minimize the risk of complete failure and at the same time prepare backup solutions that enable rapid recovery, reconstruction, and deployment. In short, they will face problems qualitatively similar to those they already face today".

2.3.1 Looking at the Future 2020

One can anticipate the following features of the future:

- The pace of technological innovation will continue to be rapid (most likely accelerating).
- The world in which technology will be deployed will be intensely globally interconnected.
- The population of individuals who are involved with or affected by technology (e.g., designers, manufacturers, distributors, users) will be increasingly diverse and multidisciplinary.
- Social, cultural, political, and economic forces will continue to shape and affect the success of technological innovation.
- The presence of technology in our everyday lives will be seamless, transparent, and more significant than ever.

2.3.2 Attributes of Engineers of 2020

The attributes of engineers would not be very different from what they are today. As per the NAE study they would be required to possess:

- Strong analytical skills.
- Practical ingenuity.
- Creativity (invention, innovation, thinking outside the box, art).
- Good communication skills.
- Familiarity with basic business and management principles.
- Leadership.
- High ethical standards and a strong sense of professionalism.
- Dynamism, agility, resilience, and flexibility.
- Lifelong learning attitude.

2.3.3 Educating the Engineers of 2020

In the follow up of the Engineers of 2020 study, another detailed study was done on "Educating the Engineers of 2020" The following quotation from the document is worthy of notice:

"This is the most exciting period in human history for science and engineering. The explosive advances in knowledge, instrumentation, communication, and computational capabilities create a mind-boggling playing field for the next Generation. . . . As we think about the plethora of challenges, it is important, in my view, to remember that students are driven by passion, curiosity, engagement, and dreams. . . . Despite our best efforts to plan their education, to a large extent we simply help to wind them up, and then step back to watch the amazing results." (Charles Vest, President MIT, July 2004)

2.4 GLOBAL ENGINEERING EXCELLENCE STUDY

To respond to the growing international concern about the preparation of engineering workforce in the future, Continental AG commissioned an international study with eight universities known for their engineering programs. Universities Participating were (1) ETH Zurich, Switzerland, (2) EPU Sao Paulo, Brazil, (3) Georgia Institute of Tech. USA, (4) MIT, USA (5) Shanghai Jiao Tong University, China (6) Technical University Darmstadt ,Germany (7) Tsinghua University, China and (8) University of Tokyo, Japan. It is easy to speak of "engineering programs" and "universities" as if they were all alike. But engineering programs within universities are as different as the systems of higher education and the national economies in which they are embedded. The global situation has a very different look depending on where one stands. Despite their many differences, the team of universities share a common goal: to provide their students with a world-class education. Furthermore, despite their diverse histories, cultures, economies, and engineering infrastructures, the team of universities shares a common vision of the need for a dramatically different kind of engineer. Together, these universities evaluated global engineering and engineering education and identified the critical factors necessary for educating tomorrow's engineering workforce. They focused on the following questions:

- Is tomorrow's engineering workforce prepared to meet these challenges?
- Are new skills required to be not only a good engineer but also a global engineer?
- Does engineering education need to change for the global age?
- Are foreign language skills and cross-cultural competence important for future engineers?
- Will globalization lead to a status gap between engineers who comfortably manoeuvre in an international environment and engineers who do not?

2.4.1 Specific Characteristics of Education System of Individual Countries

The Global Study identified the following basic characteristics of the education system of the participating countries:

- **Brazil** enjoys the advantage of engineers who are nimble at technology transfer, but the industry needs to increase its technological development capability. The country also needs a more technologically advanced infrastructure and engineers need greater social standing and compensation.
- **China**'s population offers tremendous long-term engineering human resource potential, but the near-term industrialization needs of China and its environmental concerns demand attention. While Chinese engineering graduates

today could benefit from more innovation and creativity, their strong work ethic and academic skills are highly valued around the world.

- **German** engineers enjoy high social standing and are known for their innovation, precision, and quality of work. But their lack of mobility and flexibility challenges the ability of Germany to maintain an adequate engineering workforce.

- **Japan**'s island geography has produced the world's most resource-efficient, environmentally conscious and flexible engineering workforce, but its traditional customs have also produced "workforce islands" in its industrial community. A strong commitment to one's company fosters a high-degree of teamwork and leverages seasoned engineering know-how, but it also diminishes industrial and educational mobility.

- **Switzerland** benefits from a long history of multicultural assimilation and international cooperation. While Swiss engineers are known for their strength in science and analytical and problem-solving skills, there is a need to infuse more entrepreneurship and innovation in Swiss engineering education.

- **The United States** enjoys a robust and technologically advanced economy, but it is increasingly dependent on foreign-born engineering talent. When this is coupled with the growing outsourcing of industrial R&D, American engineering faces a challenging situation.

2.4.2 Attributes of a Global Engineer

Based on the Study, the Global Engineer should be:

- technically adept;
- broadly knowledgeable;
- innovative and entrepreneurial;
- commercially savvy;
- multilingual;
- culturally aware;
- knowledgeable about world markets;
- professionally flexible and mobile.

2.4.3 The Recommendations of the Study

The Study made four major recommendations to be adopted by Institutions to achieve excellence, acceptability and mobility for their engineering graduates in the global market place.

- Global competence needs to become a key qualification of engineering graduates.
- Transnational mobility for engineering students, researchers, and professionals needs to become a priority.

- Global engineering excellence depends critically on a mutual commitment to partnerships, especially those that link engineering education to professional practice.
- Research on engineering in a global context is urgently needed.

Only a genuine commitment and sustained collaboration among all the stakeholders in engineering education will ensure a substantially increasing number of well-qualified, globally prepared engineer's world wide.

2.4.4 The Road Blocks to Global Engineering Concept in India

There are several difficulties and roadblocks in implementing the concept of Global engineering in India. Some of them include:

- Preparation for global practice is generally not viewed as central to an engineer's education.
- International mobility remains a challenge. Problems involving visas, costs, cultural barriers, language, curriculum structures, national funding restrictions and other factors make international collaboration a challenge. The Government has not yet taken a policy decision of permitting foreign engineering education providers free access to establishing universities, institutions or training organizations in India.
- Globalization and collaboration need to go hand-in-hand. Increasing the supply of global engineers will require more meaningful and substantial partnerships with global institutions committed to long-term results.
- There is a significant lack of knowledge about proven theories and effective practices for instilling global competence. Most programs are thoughtfully designed and carefully implemented, but they are seldom rigorously and scientifically evaluated for their educational impact.

2.4.4.1 *The Prevailing Aids*

In spite of the above roadblocks ,there are certain positive features which may help in the attempt to train our engineers towards global competence permitting them to work in international teams in different countries.

- Powerful and portable communication and collaboration technologies are transforming the educational environment in the country.
- Information technologies are also extending the reach of on-campus students beyond the classroom walls and are simultaneously allowing remote professionals to reach into the college campus.
- Many institutions are recognizing the pedagogical value of learning experiences beyond the formal curriculum. Cooperative education, internships, research experiences, service learning, study abroad, and similar experiential programs provide students with opportunities to better connect the theory and practice

of their fields. Hands-on learning through international opportunities also allows students to develop crucial professional skills such as teamwork and communications.

- Competence-based education is reshaping the educational experience. Competence-based curricula help sharpen an institution's educational focus, promote a more synergistic approach to education and encourage interaction with their stakeholders such as students, alumni, industry and governing boards
- India has become a provisional member of the Washington Accord and is likely to become a full signatory, which would permit equivalence of Indian qualifications among all other signatories to the Accord.

2.5 WTO, GATS AND EDUCATION

The World Trade Organization (WTO) was established by replacing the General Agreement on Trade and Tariffs at the Uruguay Round in 1994. The General Agreement on Trade in Services (GATS) covered in the WTO, also a product of the Uruguay Round, is a legally enforceable agreement aimed at deregulating international markets in services, including education. The objective of GATS is to liberalize trade in services as quickly as possible. It is clear from the preamble of GATS that it is a "multilateral framework of principles and rules for trade in services with a view to the expansion of such trade under conditions of transparency and progressive liberalization" and with a 'desire' for the "early achievement of progressively higher levels of liberalization of trade in services through successive rounds of multilateral negotiations." The WTO has defined trade in services in the following four forms of the supply of a service:

(a) *"from the territory of one member into the territory of any other member."* Called **Cross Border Supply**, this service in education includes any type of course provided through distance education, or Internet, or any type of testing service and educational materials that can cross national boundaries. When the institution of a member country A, provides distance courses, etc., to another member country B, then A is deemed to be exporting education service to B.

(b) *"in the territory of one member to the service consumer of any other Member."* Called **Consumption Abroad**, this refers to the education of foreign students. When the students of a member country A, move to another member country B, then B is said to be exporting education service to A.

(c) *"by a service supplier of one member, through commercial presence in the territory of any other member."* Called **Commercial Presence**, this refers to the actual presence of foreign supplier in a host country. This would include foreign universities or providers of a member country A, setting up courses through branches or franchisees or entire institutions in another member country B. A would be deemed to be exporting education service to B.

(d)   *"by a service supplier of one member, through presence of natural persons of a member in the territory of any other member."* Called **Presence of Natural Persons** in WTO jargon, this refers to when foreign teachers of a member country A, moving to teach in another member country B, A would be deemed to be exporting education service to B.

When the services are entirely provided by the government, they do not fall within the GATS rule. For a service to be out of the purview of the GATS rule, it has to be entirely free. However, when the services have been provided either by the government partially or some prices are charged (as happens in education where some fees is charged), or provided by the private providers, they shall fall under the GATS rule.

2.5.1 GATS in Education and India

India although a member of WTO and GATS, has so far made no commitment on including education under services but is under pressure for its inclusion from several countries (Australia, Brazil, Japan, New Zealand, Norway, Singapore, and US). Requests mostly focus on higher education, adult education, and other education services (US has specified training services and educational testing services, and Brazil has requested inclusion of primary and secondary education services). However India has both import and export interests in this sector. There are more than 80,000 Indian students studying in US Universities alone and several thousands more in UK, Central Europe, Russia, Australia, Singapore and China. Many foreign Institutions are entering India through twinning and franchise arrangements and offering foreign degrees to Indian students studying in their own country. Distance education and web-based programs are also being made available to Indian students, and a few foreign faculty and teachers come to India through many exchange programs and agreements. India's export interests include offer of distance learning, education process outsourcing with remote tutoring, attracting foreign students for its prestigious courses in engineering and medicine (currently round 8000, with capability of multiplying this many fold), setting up campuses abroad (for example-MAHE, BITS, CBSE schools, IMT, proposed IIM) and sending school teachers and university lecturers to countries e.g., West Asia, South East Asia, U.K., U.S.A and Africa. India has the **Potential to become a regional hub** for exporting higher education services. This context has led to a serious debate in the country on the desirability of opening up the higher education sector to foreign education providers. According to some experts, "the impact of opening up higher education services is shaped not by WTO but by domestic factors, including the domestic regulatory framework and the state of the domestic education system in terms of quantity, quality, costs, infrastructure and finances. In this context, evidence suggests that some of the concerns about opening up education services may not be so misplaced. While there are reputed foreign educational institutions operating, there are numerous less reputed, second or third tier ones as well that charge high fees for programs of dubious quality. Given India's capacity constraints in higher education, substandard foreign institutions are able to survive in India. Actually the problem is not liberalization per se, but the lack of a supportive domestic regulatory framework, which can ensure that liberalization is beneficial. This is not to suggest that one should add more layers of regulation in higher education. Already there is a plethora of regulatory bodies duplicating each other's functions. What is required is a regime of more effective registration and certification systems that prevent unapproved institutions from partnering in order to protect and inform consumers, enable good quality foreign institutions to enter the Indian market, and create a level playing field for domestic and foreign institutions so that the former can compete effectively in a liberalized environment. Once such a regulatory framework is in place, India need not fear scheduling education services under GATS. It could even inscribe additional conditions on the nature of foreign participation in higher education, something permitted under GATS commitment structure. Finally, a point often lost on critics is that India also has gone on the offensive in education services. A growing number of Indian educational institutions are beginning to export to other markets. So, globalization of education services should also be seen as an opportunity and the GATS as a framework to exploit this opportunity. In short, a pro-active rather than defensive approach is required to benefit from the liberalization of higher education services, both unilaterally and multilaterally, on the import as well as export fronts."

2.5.2 Concerns on Globalization of Education

Education, as a service industry, is part of globalization process under the umbrella of GATS. There is, however, a distinct possibility that this might "force countries with quite different academic needs and resources to conform to structures inevitably designed to service the interest of the most powerful academic systems and corporate educational providers... breeding inequality and dependence" (Altbach, 2002). Further, "Globalization... can lead to unregulated and poor quality higher education, with the world wide marketing of fraudulent degrees or other so-called higher education credentials…" (World Bank's Task Force, 2000). India is likely to turn into "an increasingly attractive market for foreign universities and hence other nations are going to use GATS' provisions to their advantage" (Arun Nigvekar, 2002). While these are obvious problems, globalization can also have advantages, particularly for India, which has a large educational system and infrastructure and diverse human capabilities.

2.5.3 Merits of Internationalization

Implicit in the system of Globalization is the inevitability of internationalization of the educational system, particularly at the higher education stage. This is not peculiar to India but has become a worldwide phenomenon. The entry of the World Trade Organization (WTO) and the inclusion of educational services under the general Agreement on Trade in Services (GATS) have given a boost to the internationalization of higher education. The merits of internationalization of education are:

(i)   Education will improve if it is internationalized and healthy competition takes place;

(ii) It will provide global opportunities and promote international good will; and

(iii) It will encourage exchange of scholars. This can be done by involvement of reputed scholars in the respective countries in curriculum designing and transaction of knowledge.

2.5.4 Indian Safeguards

Notwithstanding the merits of internationalization of education, it will be harmful for our country to allow an unregulated entry of foreign institutions in India. Quite a few of these institutions are of dubious quality – some of them have not even been recognized in their own countries. In a Study *"Partnership of Private Sector in Financing and Management of Higher Education" (2002).** in consultation with academics, educational planners and representatives of the industry, the following guidelines were suggested about the entry of foreign institutions in India:

- While the foreign institutions may be allowed to set up their campuses in India, they should function under the control of the government or special ized bodies like the National Assessment and Accreditation Council (NAAC) set up by the Government;

- The universities which want to function in India should have been accredited in their own countries;

- The foreign institutions should be subjected to pre-entry academic audit and accreditation norms devised by the designated government agencies;

- They should sign Memoranda of Understanding (MOU) with the government or a body designated by it. The Memorandum should give details of the courses of studies, infrastructural facilities, both academic and non-academic and the amount of expected cost recoveries from Indian students;

---

*   The Study was conducted at the instance of the Planning Commission at the National Institute of Educational Planning and Administration (NIEPA) under the direction of Prof. J L Azad.2002).

- The entry of foreign institutions should be allowed on the basis of reciprocity. The countries exporting education to India should also permit the opening of Indian university campuses in their countries. There should be provision for exchange of faculty among the various countries participating in the programme; and
- In the entire process of the entry of foreign institutions, the paramountcy of national interests should be the crucial guiding factor.

2.6 OBLIGATIONS UNDER WORLD DECLARATION ON HIGHER EDUCATION FOR THE 21ST CENTURY (PARIS 9 OCTOBER 1998)

India is a signatory to the World declaration on Higher Education under which it has "reaffirmed the right of all people to education and the right of access to higher education based on individual merit and capacity", and pledged "to take all necessary measures in order to realize the principles concerning higher education contained in the Universal Declaration of Human Rights and in the Convention against Discrimination in Education". Under this declaration, the priority actions at the National level include:

- Ensuring through legislative, political and financial framework of reform that higher education shall be accessible to all on the basis of merit and no discrimination would be accepted, no one can be excluded from higher education or its study fields, degree levels and types of institutions on grounds of race, gender, language, religion or age or because of any economic or social distinction or physical disability.
- Reinforcing the links between higher education and research.
- Developing higher education institutions to include lifelong learning approaches.
- Making efforts, where necessary, to establish close links between higher education and research institutions, taking into account the fact that education and research are two closely related elements in the establishment of knowledge
- Developing innovative schemes of collaboration between institutions of higher education and different sectors of society to ensure that higher education and research programs effectively contribute to local regional and national development, and
- Defining and implementing policies to eliminate all gender stereotyping in higher education and consolidating in all levels and in all disciplines, in which they are under-represented at present and, in particular, to enhance their active participation in decision making.

The declaration and the frame work for priority action reinforces the need for restructuring higher engineering education to make it merit-based for universal access, to bring education, research, innovation and economic development as the main focus of its mission, to develop close links with research institutions, and to eliminate all types of discrimination.

2.7 ENGINEERING EDUCATION IN A KNOWLEDGE SOCIETY

Rapid advances in Information and Communication technologies have helped nations all over the world in accessing information and knowledge created anywhere in the world and using them for the prosperity and well being of its people. They have also given a major fillip to the creation and dissemination of knowledge inside the country and using it extensively for economic, social, cultural and other human activities. **Knowledge Society** normally refers to any society where knowledge is the primary production resource instead of capital and labour and where due importance is given to the use of knowledge and information in all economic activities. Creation and utilization of knowledge have been a feature of all modern societies and have always received due importance and respect, but have remained mostly localized in the community. Current technologies have eliminated the constraints of localization and geographical proximity and have provided increased opportunities for sharing, storing and retrieving knowledge and for marketing domain specific knowledge and skills both inside the country and also across the globe.

While all knowledge is important and useful, engineering education plays a very dominant role in developing knowledge and skills which are vital to the growth and maintenance of knowledge, knowledge-based and knowledge processing industries. Apart from providing specific domain knowledge in different engineering disciplines and producing experts in computer science and engineering, and information and communication technologies, all engineering graduates irrespective of their fields of study are given education and training to acquire a reasonable level of competence in problem solving skills, software development, computer applications, modeling and simulation, and environmental impact analysis areas, which are important in creating new applications for knowledge and in marketing them both for domestic applications and for export. In recent years with the advent of the knowledge age, there has been a significant increase in enrolment in engineering programs all over the world including India in computer and information sciences, software engineering and information and communication technologies. India has taken a major lead in this area and is exporting both manpower and skills in IT to most countries in the World. To maintain its leading position in this area the engineering education system in the country has to continuously improve quality, upgrade facilities, and produce graduates with globally marketable skills.

2.8 CONCLUSION

Engineering Education is continuously evolving in all parts of the world to meet the new challenges of the emerging knowledge society and the gradual integration of the world economies. Its emphasis on quality assurance through output/ outcome based criteria and through developing attributes which would enable graduates to work in a global environment, would permeate to all countries and help restructuring engineering curricula towards increased innovation and creativity and equipping all graduates with skills of co-operative and harmonious working in multidisciplinary and multicultural teams.

REFERENCES

1. *The Future of Technology:* 2003 IEEE Fellows Technology Survey, PPP Prepared for *IEEE Spectrum* By The Response Center, December 2002 (available at IEEE Website)

2. ABET 2000-CRITERIA FOR ACCREDITING ENGINEERING PROGRAMS Effective for Evaluations During the 2002-2003 Accreditation Cycle on wards, Accreditation Board for Engineering and Technology, Inc available at Website: **http:// www.abet.org.**

3. Engineering Professors' Council (UK) Graduate Output Standard, December 2000, available at **www.epc.ac.uk/publications/standards/index.php.**

4. 'NOKIA' Perception of attributes of graduating Engineer from The EPC Graduate Output Standard document available at **www.epc.ac.uk/uploads/**output_standards/ The % 20 EPC % 20 Engineering % 20 Graduate % 20 Output % 20 Sta.

5. UNESCO Position Paper on "Higher Education in a Globalized Society "Unesco 2004

6. National Academy of Engineering:Engineers of 2020 published by National Academic Press available at NAP Website www.nap.edu and also at NAE website www.nae.edu.

7. NAE: Attributes of Engineers of 2020, published by National Academic Press, available at NAP Website and also at NAE website.

8. NAE: Educating the Engineers of 2020, published by National Academic Press available at NAP Website and also at NAE website.

9. Global Engineering Excellence Study 2006 available at **www.gee-geip.org** website.

10. Rupa Chanda: Concerns for Opening up Higher Education *The Financial Express*, December 14, 2004.

11. Altbach (2002): Knowledge and Education as International Commodities: The Collapse of the Common Good, International Higher Education, vol. 28, pp. 2-5.

12. World Bank Task Force Report 2000: Peril and Promise: Higher Education in Developing Countries.

13. Arun Nigavekar (2002).

14. *Partnership of Private Sector in Financing and Management of Higher Education" (2002).* — A study carried by Prof. J.L. Azad of NIEPA (2002).

15. World Declaration, Paris 1998: World Declaration on Higher Education for the 21st Century: Vision and Action; adopted by the World Conference on Higher education, Paris, 9 October 1998. Available at **www.unesco.org/education/educprog/wche/ declaration_eng.htm.**

3

Challenges Faced by Indian Engineering Education

"When you find you are on the side of the majority, it is time to reform."

— Mark Twain

3.1 BACKGROUND

Facilities for education in engineering, technology and management have expanded considerably since Independence. Whereas there were only 44 engineering degree level institutions in the country at the time of Independence with a total intake capacity of 2570, on 31st August 2007 there were 1668 AICTE approved engineering degree level institutions with student annual admission capacity of 653,290. In addition professional education is available annually at the first degree level in Pharmacy for 52,334 students in 854 institutions, in hotel management and catering technology for 5272 in 81 institutions, in general management (MBA/PGDBM) for 121,867 students in 1149 institutions, in master of computer applications for 70,513 in 1017 institutions, and in architecture for 4543 students in 116 institutions. At the post-graduate level in engineering there are 1983 programs in 483 institutions with an enrolment capacity of 36052 students. Much of the expansion in technical education facilities has occurred during the Ninth and Tenth Plan periods due primarily to government policy support for encouraging private investment in the field. During the Tenth Plan period, the degree level institutions went up from 1057 to 1459 with an annual intake capacity going up from 295,796 to some 550,000 an increase of some 80% in less than five years. The private sector continues to be a major player in technical education with its current share in engineering today being close to 86%. Table1 shows the growth of number of engineering degree institutions in the country during 1997-98 to 2008-2009, annual growth rate of institutions and the sanctioned intake capacity (there is a slight mismatch of data from MHRD Annual Report and Banerjee and Muley Report).

While there is no consistency in the annual growth rate of institutions, the public demand for expansion of engineering education shows no sign of abetment. The sanctioned intake capacity has nearly doubled in the last five years.

Table 3.1: Growth of Engineering Degree level Institutions

Year	No. of engineering degree institutions	Annual Percentage Increase	Sanctioned Intake
1997-98	562	—	134,894*
1998-99	644	14.6	na
1999-00	755	17.2	na
2000-01	821	8.7	na
2001-02	1057	28.7	na
2002-03	1195	13	359,721*
2003-04	1263	5.7	439,689*
2004-05	1358	7.5	459,407*
2005-06	1476	8.7	517,018*
2006-07	1522	3.1	na
2007-08	1668	9.6	653,290
2008-09	2388	43.2	820,000

Source: Annual Report MHRD 2006-07, data from AICTE and Banerjee & Muley of IITB Report—Engineering Education in India 2007.

In the last two years the demand situation appears to have gone berserk (AICTE-approved engineering degree level institutions have gone up in 2008-09 by over 40% to 2388 with the annual sanctioned intake capacity going up by more than 25% to over 820,000 students, while the new request for approval for the 2009-10 session is reported to be another 880 institutions). This sudden jump in rate of growth could be attributed to the continuing attractiveness of degree level programs for employment of graduates in the growing economic environment, but this could also be due to increasing confidence of the private sector in investing resources in engineering education for getting fairly high returns on investment. While the increase in the number of engineering institutions offering degree programs in the last few years has opened the doors for increasing the trained manpower pool for the economy, challenges of increasing access with equity, reducing regional imbalance, and improving and assuring quality of offerings continue to remain areas of major concern. These issues are discussed in some detail in the following paragraphs.

3.2 ACCESS

The accelerating growth of the Indian economy, between 8-9 % p.a., has generated increased annual demand of qualified engineers in construction, manufacturing and service sectors, so much so that employers are already reporting shortages both in the core industries as well as the IT sector. The Nasscom-Mckinsey Report (2005) made a forecast of a shortage of some 500000 engineers in the IT and ITES sector alone by 2010. When this is coupled with the demand of Indian engineers in other parts of the world, even the present high rate of expansion (average of some 14% per year during the last ten years) may prove to be inadequate. The reported decision of the Government to raise the Gross Enrolment Ratio (GER) of Higher education from the present 10% to 15% by the end of the 11th Plan and 22% by the end of the 12th Plan would in itself need a higher rate of growth of engineering education in the country during the next decade, since engineering and medicine are the preferred disciplines for admission to higher education in the Country. Although the proposed expansion and the rate of growth in Higher Education facilities are most welcome as they would help meeting the growing demands of an economy which is poised to reach the double digit rate of growth, India is still lagging far behind in the production of S&T manpower when compared to those of other countries: India having 3.5 S&T personnel per 1000 population against China having 8.1, South Korea 45.9, USA 55, Germany 76, Israel 76 and Japan 110.If India has to emerge as a developed industrialized nation, it has to expand access to professional education many fold. Compared to China, India presents a much poorer picture. In 2004, for instance, there were 4,376,167 engineering students studying in various universities and colleges in China, which constituted about 34% of the total number of university students, i.e., 13,334,969. India on the other hand in the same year had 696609 students studying Engineering in AICTE approved institutions, which was only about 6% of all Higher Education students (11,777,296). At the post-graduate level, there were 302,296 graduate students (Master's and Doctoral candidates) studying engineering at universities in China, which is 38.84% of the total number of graduate students in that country (779,408), compared to this India had some 28,000 students studying post-graduate programs in engineering which is only about 3-4% of all post-graduate students (872,161). This dichotomy shows the low emphasis placed on engineering education both at the degree and post-graduate levels in India compared to that in China .If the Indian economy has to catch up with the Chinese economy, the percentage of students entering engineering education must increase substantially.

3.2.1 Equity

In addition to the need for increasing access to engineering education to a larger number of students graduating from senior secondary schools (+2 schools), there is also a major question of equity, most of the engineering students come from affluent urban families and equal opportunities for seeking professional education are denied to those coming from rural homes, from low-income families, or to those who come from disadvantaged groups-women, scheduled castes, scheduled tribes and other backward classes. Although there have been reservations for seats in public institutions for SC/ST candidates for a long time and scholarships are awarded to all students of this category admitted to these institutions to meet their educational and living expenses, the participation of this group in engineering is still very limited when the whole of the engineering education system is considered (there is presently no reservation in private institutions which control 86% of the enrolment) The recent decision of the Government to extend the reservation benefits to other backward classes in central institutions is intended to improve equity. However the question of assisting eligible students from poorer families who do not have access to support through grant of soft loans to be recovered after their employment continues to be a problem to be tackled much more purposefully than being done at present. Some selected data from SES2005 given below illustrate the equity issue. The data (Table 3.2) refer to the entire Higher Education Sector in the absence of similar data for the engineering sector alone).

Table 3.2

Category	Total	Male	Female	Scheduled Caste	Scheduled Tribe
Total enrolment in engineering 2004-05	696609	531207 (76.3%)	165402 (23.7%)	59299 (T) (8.5%) 47076(M) 12223(F)	21507(T) (3.1%) 17288(M) 4219(F)
GER in Higher Education 2004-05	10.0	11.6	8.2	6.7(T) 8.1(M) 5.2(F)	4.9(T) 6.3(M) 3.5(F)
Gender Parity in Higher Education	0.71			0.64	0.55
Number of Engineering students Who Graduated in 2003	127610	101143 (79.3%)	26467 (20.7%)	na	na
Number of Engineering Post-graduates in 2003	12370	10205 (82.5%)	2165 (17.5%)	na	na

The gender parity in engineering education (0.31) is much worse than in total higher education (0.71). This is true not only for all students but also for the SC (0.26 as against 0.64) and ST students (0.24 against 0.55). The situation has however started improving in recent years due to a large number of girls entering computer science, information technology, electronics and communication and biotechnology programs in engineering. Apart from gender disparity, there are differences in access to higher education on the basis of religion, rural/urban divide as well as poor/non-poor divide because of perceived cultural preferences, awareness of employment opportunities, and cost of education. National Sample Survey data (NSS2003) for GER for Higher Education illustrate these variations: Total 13.2, Male 15.3, Female 11.0; SC 7.5, ST 5.0,OBC 11.34, others 24.89; Hindu 12.0, Muslim 8.2, other religions 30.9; Rural 7.8, Urban 27.2; Poor 2.43, Non-poor 12.81. This difference for higher education in engineering could be even more pronounced since the cost of professional education is relatively much higher than general higher education and girls in most communities do not prefer engineering as an attractive career option.

3.3 REGIONAL IMBALANCE

Unfortunately, the recent expansion of facilities has created not only the problem of maintaining high quality of education, it has also increased regional imbalance of educational facilities. **The facilities for engineering education is so badly distributed in the country- that 63.6% of the UG seats and 62% of the PG Seats are available in only three of the seven regions of the country namely the Southern, the South Western and the Western (Table 3.3). Further, nearly 60%of engineering graduates and post-graduates come from only 4 states, Tamilnadu, Andhra Pradesh, Karnataka and Maharashtra (Table 3).** Table 3.3 shows in detail the regional availability of sanctioned UG and PG seats.

Table 3.3: Regional Distribution of Sanctioned Intake for UG and
PG Programs in Engineering

Regions (as per AICTE groupings)	*Under Graduate Programs in Engineering as on 31.08.2007*		*Post-graduate Programs in Engineering as in Session 2007-08*	
	Number of Institutions	Sanctioned Intake Capacity	Number of Institutions	Sanctioned Intake Capacity
Central (MP, Chattisgarh, Gujarat	166	66161 (10.13%)	46	4334 (12.02%)

(*Contd.*)

Regions (as per AICTE groupings)	Under Graduate Programs in Engineering as on 31.08.2007		Post-graduate Programs in Engineering as in Session 2007-08	
Eastern (Mizoram, Sikkim, Orissa, W Bengal Tripura, Meghalaya, Arunachal Pradesh, A&N, Assam, Manipur Nagaland, Jharkhand).	128	40613 (06.22%)	37	2686 (07.45%)
Northern (Bihar, UP, Utaranchal)	155	57988 (08.88%)	34	2100 (05.83%)
North West (Chandigarh, Haryana, J&K, Delhi, Punjab, Rajasthan, H.P.	206	73251 (11.21%)	50	4563 (12.66%)
Southern (Andhra Pradesh, Pondicherry, Tamilnadu)	593	256572 (39.26%)	193	14081 (39.06%)
South - West (Karnataka, Kerala)	234	91939 (14.07%)	56	4545 (12.61%)
Western (Maharashtra, Goa, Daman & Diu, Dadar NH)	186	66767	67 (10.22%)	3743 10.38%)
INDIA	1668	653290 (100.%)	483	36052 (100.%)

Table 3.4 gives the sanctioned seats for degree and post-graduate education in engineering per lakh of population of the 30 States and Union Territories. The variation is very large: no engineering degree education is available in the north eastern states of Nagaland, and Mizoram, and no post graduate engineering education facilities in these States and in Meghalaya, Manipur and Tripura; the range for UG seats per lakh of population in the remaining States and Union Territories vary from the lowest of 1.86 for Bihar to 189.20 for Tamilnadu. Twenty States/UTs have below the national average of 58.74 UG seats per lakh of population and only 10 are ahead with Karnataka, Andhra Pradesh and UTs of Chandigarh/Pondicherry having more than twice the national average. The status of PG facilities is similar, varying from a low of 0.2 for Bihar to a high of 10.94 for Tamilnadu and 17.51 for the UTs of Chandigarh/ Pondicherry.

Table 3.4: State wise Sanctioned Seats per Lakh of Population in Degree & PG level courses in Engineering

Sl. No.	States/UTs	Projected Population, 2006	Degree Level As on 31.8.07		Post Graduate Level as in 2007-08 Session	No. Per one Lakh Population UG + PG	
1.	Nagaland	21.19	0	0	0	0	0
2.	Mizoram	9.46	0	0	0	0	0
3.	Bihar	907.52	1690	1.86	180	0.2	2.06
4.	Assam	286.65	870	3.03	156	0.54	3.57
5.	Manipur	23.08	115	4.98	0	0	4.98
6.	Meghalaya	24.70	240	9.71	0	0	9.71
7.	J & K	109.41	1401	12.81	36	0.33	13.34
8.	Tripura	34.07	490	14.37	0	0	14.37
9.	Jharkhand	292.99	3438	11.73	888	3.03	14.76
10.	West Bengal	852.16	16968	19.91	1318	1.55	21.46
11.	Arunachal Pradesh	11.69	210	17.96	54	4.62	22.58
12.	Uttar Pradesh	1,832.82	51775	28.25	1669	0.91	29.16
13.	Himachal Pradesh	64.55	1807	27.99	162	2.51	30.50
14.	Chattisgarh	225.94	7006	31.01	356	1.58	32.59
15.	Gujarat	549.79	17408	31.66	1515	2.75	34.41
16.	Rajasthan	622.76	20683	33.21	804	1.29	34.50
17.	Orissa	388.87	17817	45.82	270	0.69	46.51
18.	Delhi	160.21	6943	43.34	818	5.11	48.45
19.	Uttranchal	92.19	4523	49.06	251	2.72	51.7
20.	Goa	14.92	809	54.22	69	4.62	58.84
	INDIA	11121.86	653290	58.74	36052	3.42	62.16
21	Maharashtra	1,048.04	65958	62.93	3674	3.50	66.43
22	Madya Pradesh	663.90	41747	62.88	2463	3.70	66.58
23	Haryana	233.14	22750	68.51	1125	4.83	73.34
24	Punjab	260.59	18879	72.45	1325	5.08	77.53
25	Sikkim	5.76	465	80.73	0	0	80.73
26	Kerala	332.65	29790	89.55	1061	3.19	92.74
27	Karnataka	562.58	62149	110.47	3484	6.19	116.66
28	Other UTs (Chandigarh, Pondicherry)	31.76	3458 (788+ 2670	108.88	556 (294+ 263)	17.51	126.39
29	Andhra Pradesh	807.12	130669	161.90	6693	8.29	170.19
30	Tamil Nadu	651.35	123232	189.19	7125	10.94	200.13

Note: Figures above do not include IITs, and NITs (not covered under AICTE)

On the basis of degree level opportunities for engineering education given in Table 3.4 one can divide all the states/ UTs in five categories as given in Table 3.5.

Table 3.5

No.	Category Seats	On the Basis of availability of Engineering Degree level
1.	Those up to 0% of national average	Nagaland, Mizoram, Bihar, Assam, Manipur, Meghalaya, and Jharkhand
2.	Those between 21-40% of national average	Jammu and Kashmir, Tripura, Arunachal Pradesh, and West Bengal,
3.	Those between 41-60% of national average	Himachal Pradesh, Uttar Pradesh, Chattisgarh, Gujarat and Rajasthan,
4.	Those between 61-100 of national average	Orissa, Delhi, Uttaranchal and Goa,
5.	Those above the national average	Madhya pradesh, Mahrashtra, Haryana, Punjab Sikkim, Kerala, Karnataka, Chandigarh/ Pondicherry, Andhra pradesh, Tamil Nadu.

The above regional disparity would not get removed or even lessened if left entirely to market forces and private sector initiatives, since the latter are usually oblivious of social concerns and invest mostly in states where the atmosphere is conducive and return on investment is high. Although regional imbalance has been a concern for AICTE and the Government for many years there is no apparent radical redressal. Table 3.6 which gives Zone-wise distribution of institutions and intake capacity over a four year period (2004-2008)shows that left to market forces the regional imbalance would not get reduced significantly. Although Central and Northern zones have picked up some steam , Eastern region as expected is still struggling both in the number of institutions and the intake capacity. Even in States where the number of institutions have increased in recent years, most of the new institutions have come up not only in urban areas but in the same cities where adequate facilities for engineering degree level education already exists.

Table 3.6: AICTE Zone wise growth of the Number of Engineering Degree Institutions(NOI) and sanctioned annual intake capacity

Zone	2004-04 NOI Intake		2005-06 NOI intake		2006-07 NOI intake		2006-8 NOI intake		% change in 4 years NOI Intake	
Central	112	37195	115	39434	146	49238	166	66161	48.2 (10.13%)	77.9
Eastern	114	34016	110	33343	124	36437	128	40613	12.3 (6.22%)	19.4
Northern	106	32298	107	34748	141	43613	155	57988	46.2 (8.88%)	79.5
North-Western	153	50645	155	51295	183	58304	206	73251	34.6 (11.21%)	44.6
Southern	496	165757	502	178637	534	196013	593	256571	19.6 (39.26%)	54.8
South-Western	207	70788	206	72063	221	77972	234	91939	13.0 (14.07%)	29.9
Western	158	48990	160	49867	173	55441	186	66767	17.7 (10.02%)	36.3
All India	1346	439689	1355	459407	1522	517018	1668	653290 (100%)	23.9	48.6

Hence there is an urgent need for some innovative State interventions not only to reduce this growing disparity but also to promote location of new institutions in educationally deprived areas in the State. The five categories of States mentioned earlier need separate strategies to reduce the imbalance. Category 5 States need no State intervention and increase or decrease in availability can be left to market forces. AICTE should reduce the number of sanctioned seats if seats remain unfilled for two consequitive years, whereas the States can insist that new technical institutions would be permitted only on the assessed manpower demand. For other categories, AICTE in consultation with the concerned States should prescribe targets to be reached in the next three Plan periods. It appears reasonable that Category-1 States should reach seat availability of 40% of the current national norms in the 11th Plan period, and 60% by the end of the 12th Plan. Category 2 states should reach a target of 60 percent in the 11th Plan period and 80% by the end of the 12th Plan; Category 3 States should reach 80% of the current national norms by the end of the 11th Plan period, and 100% by the end of the 12th Plan period. Category 4 states should reach the current national norms by the end of the current plan period. Although during the 11th and 12th Plan periods, the states which are above the national average would continue to start more new institutions and the national norms will change, the planning should be done on present norms and reviewed at the end of the 11ᵗʰ Plan. The specific interventions would vary from State to State even in the same category and would need detailed discussion between AICTE and the State authorities As already mentioned earlier that even in States/UTs which have adequate facilities for degree level engineering education, most of the institutions are located in urban areas with very limited access to rural students. Category 4 and 5 States should in future try to reduce the district-wise regional imbalance in the States by encouraging starting of new institutions in so far neglected districts. Some common strategies, which can be followed are listed in the Table 3.7.

Table 3.7: Strategy to Reduce Regional Imbalance

State Categoris	*Targets*	*AICTE Initiatives*	*Central Initiatives*	*State Initiatives*
Category 1	Double facili- in each State and reach 40% of national norm at the end of plan.	Increase number of seats in existing institutions provi- ding one time fixed grant to state institutions; provide 50% Capital assistance to new State insti- tions; provide 50% Capital assistance to new State insti- tutions and a fixed grant to private institutions.	Set up new central institu- tions in appro- priate disciplines; provide adequate plan funds for states in techni- cal education.	Indentify place for setting up new institut- tions, give free or subsidized land to private investors in technical edu- cation, provide other incenti- ves.
Category 2	To reach 60 % of national the end of of plan.	Increase number of seats in exist- grant to State institutions, pro- vide 30% capital assistace to new State Institutions, provide 30% cap- ital assistance to new state Institu- tions and a fixed grant to private instituions.	Set up a few Central Insti- identified disciplines, provide ade- quate plan funds to states for exp- ansion of techni- cal education.	Invite private investments sion of facilit- ies through in- centives, set up state institut- ons in areas of high demand.

(Contd.)

State Categoris	Targets	AICTE Initiatives	Central Initiatives	State Initiatives
Category 3	Reach 80% of national norms by end of plan	Increase number of seats in existing institutions; Provide 20% capital assistance to new state instititons and a fixed sum to private institutions.	Provide adequate plan funds to states, bring good institutions under national quality improvemnent program.	Improve State level planning and quality assurance, provide incentives for private investments.
Category 4	To reach 100% of national norms by end of plan	Increase seats in existing institutions and support new state institutions and support new state institutions through a fixed subsidy, provide some incentives to private institutions.	Provide adequate plan funds for technical education, bring selected ones under national quality improvement program.	Support private initiatives through incentives. Identify support to existing and new state institutions.
Category 5	Improving Quality; All new programs only on detailed manpower study and preferably in neglected disticts	Reduce number of seats in courses remaining unfilled; Support quality through incentives	Bring selected ones under national quality improvement program for quality up gradation; no new central institutions.	Start Quality Assurance cells in all institutions and have a state level monitoring cell.

While it is not possible to drastically reduce the imbalance of technical education facilities in one plan period, much could be achieved by a declared policy of reducing this imbalance through pro-active State intervention for which a special fund needs to be created in each plan period for the 10th, 11th and 12th plan, and the programme of implementation reviewed after each plan.

3.4 QUALITY

The recent large-scale expansion in engineering education has come at the cost of the quality of the educational offerings. The affiliating University system, outdated curricula, inadequate academic infrastructure, shortage of qualified teachers, poor teaching/learning processes, lack of innovative and creative activities, archaic evaluation system, absence of proper academic ambience, and non-participative and bureaucratic governance mechanisms have all contributed in different degrees to the lowering of standard of the educational offerings of a large proportion of the 1700 engineering institutions in the country. Apart from the IITs, the IISc and a few well renowned engineering colleges and University institutions that are comparable in quality to the very best in the world, most of the engineering institutions, both in the public and the private sectors, are facing enormous problems in satisfying international quality standards. The affiliating university system kills all faculty initiative of improving and modernizing curricula to meet emerging market demands. Shortage of teachers coupled with the long time taken in filling faculty vacancies makes the contact load of teachers so high that they have no time for introducing innovation in the teaching/learning processes or to indulge in creative research and development activities. The situation is so bad that a recent Nasscom-Mckinsey report (2005) has mentioned that only 25-30% of engineering graduates in India are employable in the IT/ITES sector However the growing demand of the economy has compelled the Engineering Industry to recruit even the low quality output from the education system and makeup the deficiency in education and training through specialized training programs which may last from 12 to 18 months. Some enterprising high-tech firms have come up with innovative schemes through which they influence both the curricula and competency level of teachers of selected institutions and thus improve the competency level of the graduating engineers. The Government is in a catch-22 situation where it must expand engineering education to meet anticipated shortages and yet keep the output standards at a minimum acceptable level even when trained teachers are not readily available. Large-scale training of existing engineering teachers and improving the academic infrastructure were given high priority in the 10th Five-year Plan and are likely to continue in the 11th and 12th plan periods along with new incentives to graduates for doing post-graduate work and entering the teaching profession. Improvement of the quality of educational offerings would need concentrated effort not only on autonomy and training of teachers but also on improving curricula and course contents, motivating teachers and students towards innovative and creative activities, equipping students with the four important skills of problem solving, design, self-learning and communication, and developing active interaction with industry and community.

3.5 QUALITY ASSURANCE, ACCREDITATION AND ISO CERTIFICATION

While there are several quality assurance and quality management systems available in commerce and industry, accreditation of educational program offerings by independent agencies is recognized as the most valuable tool for quality monitoring and certification in the educational system. Before the establishment of Accreditation agencies, the Universities themselves through an internal mechanism, which included among others the following, assured the responsibility of ensuring the quality of education imparted by Institutions/Departments under the jurisdiction of Universities:

- Visit to the Institutions by an expert team for ensuring adequate facilities for imparting education.
- Review of proposal by a Peer group.
- Curriculum control through University Academic Council.
- Performance evaluation of students through University Examinations.
- Prescribing the qualifications for teachers.
- Ensuring through regulations the educational attainments of students at the time of admission.
- Linking grant of public funds to institutional performance.

The Universities made mandatory inspections before the starting of the programmes, during the running of the programmes and at subsequent intervals as deemed appropriate. With the growth in numbers of Universities running similar programs, it became desirable to ensure that all programs satisfied at least some minimum quality norms. Such assurance would permit acceptance of degrees awarded by one University by another and also help employers to assess the market value of the University degrees. This requirement led to the establishment of coordinating agencies/ accreditation councils, which laid down specific and transparent criteria, which must be satisfied by all institutions for ensuring quality of the educational process. In India, the University Grants Commission, the All India Council for Technical Education and the Association of Indian Universities performed the major co-ordination role for ensuring equivalence of achievement of similar University Degrees awarded by different Universities in India and abroad. Later two autonomous bodies the NAAC and NBA were established to accredit programs in general higher education and engineering education respectively. Both the NAAC and NBA used the well-tried detailed input criteria to grant accreditation of programs. These were based on the then internationally accepted norms which included faculty qualifications, student/teacher ratio, adequate infrastructure facilities in terms of classrooms, laboratories, equipment, library, sports and cultural activities, hostel accommodation, teaching/learning methodologies, relevant and effective up-to-date curricula, a transparent and fair evaluation mechanism, and a supportive management. The main difference between NAAC accreditation and that of the NBA is that while the former is Institution based certifying all programs of the University/Institutions, the NBA accredits individual program offerings of an institution.

The present NBA accreditation process with which most of engineering institutions are concerned, suffers from many inadequacies, which include:

- Failure to cope up with the requirements of accreditation of a very large system. Inadequate NBA staff, large processing time, not very effective use of online processing of data, and slow pace of decision making are all contributing to the pile up of backlog of programs to be accredited.
- Output oriented criteria for accreditation has not yet been fully developed yet and implemented. Abilities in knowledge, skills, and attitudes that accredited programs must ensure in their graduates, need detailed specifications, circulation and implementation
- Very little concern for learning outcomes.
- Accreditation remaining voluntary, poor quality institutions do not usually apply for accreditation for fear of being refused accreditation of a program offerings. As of January 2008, only some 2000 engineering degree programs were accredited which is less than 40% of the running programs.

With the increase in delivery mode the process of accreditation has become very demanding. The delivery modes in operation are:

- Traditional face-to-face interaction in a classroom
- Distance learning mode (both synchronous and asynchronous)
- On-line learning mode (asynchronous)

Each of the above delivery modes would require different quality assurance mechanism since they have different in-built capabilities as seen in the Table 3.8.

Table 3.8

Attributes	Face to face	Distance Mode	On- line Mode	Comment
Cuurricula	Adaptable to suit level of audience	Fixed due to non- homogen- ous audience	Fixed due to no-control on audience	Curricula in distance and on-line modes to be evaluated by clients them selves.
Interactivity	High and inexpensive	Low and expensive	Low and expensive	Electronic interaction in distance and on line modes
Problem solving	Direct contact	Through e-mail	Through e-mail	Initiative totally of stu- dents in D and Ol modes

(Contd.)

Attributes	Face to face	Distance Mode	On-line Mode	Comment
Group Working	High	Low	Very Low	Can improve in D mode through study centers
Communications skill development	High through assignments, colloquia, face to face interactions	Low except when using study centres	Very low	New methodologies needed to develop in OL mode.
Applications skills	High through graded assignments	High through graded assignments	Low unless specially catered for	Needs interactive practice
Creativity and innovative ability development	Teacher's initiative and encouragement	Needs special effort	Difficulty to manage	Needs new methodologies in D and OL modes
Quality Assurance	Established Porcedures	Needs research and new Procedures	Needs research and new procedures	It is still a very area for D and OL modes

In higher education in general and professional education in particular, the graduates have to acquire knowledge, specific skills and attitudes, treatment of which could be very different in the three modes as shown in the Table 3.9.

Table 3.9

Attrubutes	Face-to-Face	Distance Mode	On-line Mode	Comments
(a) Knowledge				
● Transfer	Good	Good	Good	Check and
● Assinilation	High	Low	Low	monitoring
● Application	High	Low	Low	difficult in D
● Integration	High	Low	Low	and OL
(b) Skills				
● Training	Yes	Possible	Possible	Requires
● Repective Practice, (Problem solving,	Yes	Possible	Possible	Special efforts and
design, communication, self-leaning)	Yes	Possible	Possible	procedures in Distance and OL modes
● Testing Competence				

Attrubutes	Face- to-Face	Distance Mode	On -line Mode	Comments
(c) Attitudes				
● Life-long learning	Yes	Yes	Yes	Attitude development
● Group working	Yes	Possible if study center	Not possible	is difficult in distance and OL modes
● Interdisciplinary working	Yes	Possible do	Not possible	
● Managing Diversity	Yes	Difficult	Not possible	

Recent growth of offerings of education and training by a large number of international players has created new problems in many countries including India. Some of these are:

● Course offerings in distance and O/L modes are usually for non-degree qualifications.

● Many course offerings are not accredited in the provider's own country.

● No efforts for adaptation to local ethos are made.

● Usually commercial interest dominates offerings.

● Twinning arrangements do not always optimize strength of partners. Quality control is only on paper. Exposure to different cultures and managing diversity are only one sided as student movement is in one direction only.

● On-line learning if not suitably adapted could create problems of understanding accents, phraseology, local examples etc. Reliability of outcome evaluation difficult.

● Course-wares on public domain may be very useful to teachers but not so much to students who need guidance.

● Students learn a lot laterally from colleagues, from environment and from casual discussions over coffee tables. This is lost in Distance and OL modes

● Foreign offerings do not train for managing diversity.

● Many International Universities have no Quality Assurance Mechanisms.

In the globalization scenario with WTO countries agreeing to free trade in products and services, educational offerings would increase dramatically over the next decade. To ensure that there is some understanding of the quality and value of various programs, there is a need for the International Community to regulate Quality Standards.

The following recommendations are worthy of serious consideration for regulating the entry of international education providers in India.

- All course offerings outside the provider's own country must have been accredited in the provider's own country and should obtain international accreditation and permission for adaptation when offered abroad. The accreditation agency of the host country (NBA) should be consulted for adaptation of the original course material.
- Special tools be developed to improve interactivity in distance and O/L modes of delivery.
- Proper methodologies be developed for evaluating outcomes in distance and online learning modes.
- Training providers to ensure competence building in training modules through computer assisted instruction and evaluation.
- Combination of delivery modes be attempted as a strategy to improve teaching learning processes, the advantage of any-time any- place self- learning of the O/L mode be combined with the group work, interdisciplinary activities, interactivity and group problem solving advantages of the face-to-face mode, and perhaps with the repeated exposure to recorded programs of the distance mode.
- Research be intensified in developing reliable indicators for quality assurance of the Distance and O/L modes so that credit transfers between modes become common.
- Development of the right learning and working attitudes be promoted in the asynchronous Distance and Online modes through electronic bulletin boards, chat sessions, critical reviews, email, video conferencing etc.

The Accreditation issue has become more important now that NBA-AICTE is a provisional member of the Washington Accord, which permits member countries to give mutual recognition of their degree level engineering qualifications. The NBA while becoming a provisional member has agreed to bring its accreditation policy and procedures to fall in-line with those of other countries. Once the NBA implements fully its reported output/outcome based criteria for accreditation and brings all engineering institutions under its umbrella, it would become eligible for full membership of the accord which would in effect give an international accreditation to NBA accredited programs. Strengthening the NBA, making it fully autonomous and linking quality improvement incentives to accreditation would bring a sea change on the quality horizon of the Indian Engineering Education system.

Among other management tools used for ensuring quality management in Higher Education, ISO 9000 series, Baldridge National Quality Award (BNQA), and the European Foundation for Quality Management (EFQM) have often been adapted and used for the Higher Education System. There has been some debate on the utility of ISO 9001 -2000 certification for higher education, but more and more institutions are applying for such a certification to ensure that institutions adhere to normal quality standards in all its processes similar to those of a manufacturing organization. ISO has published specific guidelines for organizations in the education sector wishing to benefit from the implementation of ISO 9001:2000, the latest, improved version of ISO's quality management system standards that have been adopted by more than half a million users in all sectors worldwide. While ISO 9000 users already include educational establishments of all types, the guidelines will facilitate implementation by the sector of ISO 9001:2000. The guidelines are intended for organizations at all levels, providing all types of education, including: elementary, medium and higher; special and adult education; distance and e learning. (ISO/IWA 2: *Quality management systems - Guidelines for the application of ISO 9001:2000 in education*, contains the full text of ISO 9001:2000, clause-by-clause, followed by specific text making the standard easier to understand and implement by the education sector. (Source: ISO News 29 Jan. 2004 ref.889)

The United States Congress enacted in 1987 the Malcolm Baldrige National Quality Improvement Act with the objective of encouraging American Business and other organizations to practice effective quality control in the provision of their goods and services, and established the Malcolm Baldrige National Quality Award Program The program was subsequently adapted to suit the educational services sector. "The Baldridge National Quality Award program is **a** powerful management system because it challenges organizations to identify and recognize existing systems within their organization. Systems encompass every aspect of any organization from student recruitment to delivery of instruction, from planning to human resource management. Once systems are documented, work can begin in continuous improvement on an organization-wide scale". (Sorenson et.al. 2005)

The EFQM Excellence Model was introduced at the beginning of 1992 as the framework for assessing organizations for the European Quality Award. It is now the most widely used organizational framework in Europe and it has become the basis for the majority of national and regional Quality Awards. The EFQM Excellence Model is a practical tool that can be used in a number of different ways: as a tool for **Self-Assessment, as** a way to **Benchmark** with other organizations, as a guide to identify areas for **Improvement**, as the basis for a common **Vocabulary** and a way of thinking and as a **Structure** for the organization's management system. The EFQM Excellence Model is a non-prescriptive framework based on 9 criteria. Five of these are 'Enablers' and four are 'Results'. The 'Enabler' criteria cover what an organization does. The 'Results' criteria cover what an organization achieves. 'Results' are caused by 'Enablers' and' Enablers' are improved using feedback from 'Results'. The Model, which recognizes there are many approaches to achieving sustainable excellence in all aspects of performance, is based on the premise that: *Excellent results with respect to Performance, Customers, People and Society are achieved through Leadership driving Policy and*

Strategy, that is delivered through People, Partnerships and Resources, and Processes. (From EFQM Bulletin).

Indian Institutions have been very slow in the use of applying management quality standards for their educational processes and systems. However there has been some movement in this regard in the last few years when many engineering polytechnics and engineering colleges have obtained ISO 9001 certification. This has enabled these institutions to declare their quality policy and philosophy, document all their processes, rules and regulations from admission, recruitment, promotion, curricula, purchase, contract, to consultancy and industry interactions, and develop a system for continuous improvement to achieve excellence and stakeholder satisfaction. This movement must accelerate and cover more engineering Institutions in the country. The AICTE should develop a National Educational Excellence model by incorporating the main thrusts of ISO, Baldrige and EFQM models and propagating the same for adoption by all Technical Institutions in the country. It could also institute National Quality Awards on the lines of the Rajiv Gandhi National Quality Award, instituted by the Bureau of Indian Standards in 1991, to encourage well performing institutions to achieve excellence in their educational offerings.

3.6 CONCLUSION

The Government, the AICTE and the engineering education community in the Country are fully aware of the challenges outlined above and some proactive action has already been taken in the 11th Plan formulation. While access and equity issues can be handled through allocation of sizeable funds and building physical infrastructure where desirable, the quality issue would need major effort over at least the next three plan periods as this would involve competence development of existing teachers, attracting through suitable incentives a larger number of engineering graduates and post-graduates to a teaching and research career, synergizing the innovation efforts and potential of educational institutions, research laboratories and industry, and above all a change in the mindset of educational managers and policy planners in removing barricades to achieving excellence.

REFERENCES

1. Annual Report Ministry of Human Resource Development, Government of India 2006-2007, and 2007-2008.
2. Banerjee and Muley: Engineering Education in India 2007, IIT Bombay Report sponsored by Observer Research Foundation
3. C.S. Jha: Higher Education in India-Restructuring for increased innovation, Document prepared for the World Bank, June 2006.
4. Selected Educational Statistics, Governemtn of India, SES 2005.

5. C.S. Jha: A Proposal for Reducing Regional Imbalance in Technical Education Opportunities; Prepared for AICTE in 2006.

6. ISO/IWA 2: *Quality management systems - Guidelines for the application of ISO 9001:2000 in Education.*

7. Malcolm Balridge National Quality Improvement Act 1987, US Congress.

8. EFQM Excellence Model 1992, details available at Wikipedia or *www.efqm.org*.

9. Lok Sabha : Information Based on Annexure referred to in parts (a) and (b)of the reply to the Lok Sabha Unstarred Question No.56 for 17.2.2009 asked by Shri Jivabhai Ambalal Patel, Shri V.K.Thummar & Shri Subhash Maharia, M.P.s regarding Setting up of Technical Institutions.

4

Ethics in Engineering Education

4.1 WHAT IS ETHICS?

A dictionary definition of ethics is that it is a science of morals, or a set of principles with a sense of purpose. Distinction between right and wrong, virtuous thoughts and actions, and being good and to maximize goodness for the greatest numbers (Gandhi Talisman) is how morality or values are defined by spiritual philosophers. Ethics is a higher philosophical pursuit, above morality and legality since what is illegal is also immoral and unethical, but not *vice versa*.

Ethics are fundamental to the civilization of any society. Starting from the dawn of civilization with hunter gatherers, ethical norms and standards have evolved in different civilizations, cultures and societies at different times, with a profound influence of culture, religion, and spiritual and philosophical beliefs. Ethics are relative, not absolute ; some having converged to a core principle of a secular and commonsense-value code of conduct, have become universal and fundamental to humanity. Beyond the fundamental ethics, there are others such as personal ethics based on individual's beliefs, interpersonal ethics developed between individuals or groups of individuals, societal ethics evolved between groups of individuals of different societies/cultures/religions/professions, professional ethics based on the practice of a particular profession, and regulatory ethics for regulating professional ethics.

The earliest known example of a code of ethics is the Hippocratic Oath for medical profession which was created during the 3rd century BC and continues to be administered even today world over, testifying to its relevance to the profession. Professional ethics continue to evolve with changing social ecologies but are specific to a profession as practiced in a particular environment. An ethical professional is expected to be an ethical individual though the two may in some cases be in conflict with each other.

4.2 WHY ETHICS FOR ENGINEERS?

Paradigmatic shifts have taken place as a result of the industrial revolution leading to the evolution and definition of ethical values and codes of conduct in business, management, scientific, engineering and other professional activities in the industrialized western world. Such ethics have gradually been adopted as an integral part of teaching-learning process in schools of management all over the world. Engineers are expected to apply science and technology for progress and prosperity of society at large. Engineers interface with society in a multidimensional way, depending on their role as a teacher, professional worker or a manager. Consequently, it is paramount for him/her to imbibe professional ethics. It is equally important to make sure that professional education also imparts such ethics to the students aiming at knowledge-based professional careers.

Today, Knowledge is the engine of growth and is the driving force behind rapidly changing, shifting and emerging paradigms in a globalised professional engineering environment. The key components of these paradigms are:

- Trade, markets and world trade organization (WTO) regime.
- Knowledge based and innovation driven technologies and manufacturing.
- Global and local competitiveness.
- Concern for ecology, sustainability, green industry, green energy and green jobs.
- Intellectual property rights (IPR) regime.
- Outsourcing & offshoring.
- Quality, reliability and total quality management (TQM).
- New, open, flat, virtual, cross cultural, and ethical management modes and styles.

Consequently, Knowledge is now a multidimensional creativity/innovation in sciences (including social sciences), engineering, technologies, manufacturing, marketing, management, etc. The evolving tech-nomic-globalisation demands adherence to internationally acceptable ethical values and code of conduct. It is, therefore, essential for academic institutions to develop the mindset of engineering students to cultivate a sense of social responsibility to uphold the honour and dignity of the profession in a global environment.

4.3 ETHICAL ISSUES

Ethical issues and responsibilities relate to functions of an individual in different professional environments of a student, teacher, professional, consultant, entrepreneur, manager, or industrialist. Issues relevant to engineering profession are:

4.3.1 Personal Ethical Issues

At personal level, issues are fundamental in nature and relate to such desirable characteristics as high levels of integrity and honesty, wisdom, loyalty, fairness, impartiality, trustworthiness, reliability, courage, compassion, humility, divinity, love and being not submissive observer or remaining indifferent to wrong happenings.

4.3.2 Academic Ethical Issues

In an academic environ, issues relate to scientific values, scientific temper, research integrity, and scientific misconduct. Based on the report of a high level committee, President Clinton gave his "Presidential Finding" on research misconduct as "fabrication, falsification, or plagiarism in proposing, performing, or reviewing research results. Fabrication is making up data or results and recording or reporting them. Falsification is manipulating research materials, equipment, or processes, or changing or omitting data or results such that the research is not accurately represented in the research record. Plagiarism is the appropriation of another person's ideas, processes, results or words without giving appropriate credit. Research misconduct does not include honest error or differences of opinion".

Besides issues related to scientific misconduct in the form of plagiarism, duplicate or recycled publications, quality and integrity of data, there are other ethical dimensions such as :transparency, accountability, quality of peer review, author credit, intellectual property rights, awards and rewards, conflict of interest, quality of research supervision, recruitment and assessment, sharing of R&D facilities, concern for sustainability and ecology, interaction with public and media, policy matters, regional, communal and caste factors, political interventions, corruption, antiquated governance and rules of management, whistle blower's role and fate, etc.

4.3.3 Professional Ethical Issues

"A Profession is a calling requiring specialized knowledge, and often long and intensive preparation, including instruction in skills and methods, as well as in the scientific, historical, or scholarly principles underlying such skills and methods, maintaining by force of organization or concerted opinion, high standards of achievement and conduct, and committing its members to continued study and to a kind of work which has for its prime purpose the rendering of public service" (Webster's Third New International Dictionary)

A professional is expected to contribute to the progress and well-being of society in a sustainable environment through creation and dissemination of knowledge, and by using engineering knowledge and skills to develop technologies, products and services, as also regulatory mechanisms, etc consistent with an accepted ethical code of conduct.

In addition to personal and academic issues mentioned in the proceeding, a professional faces a whole variety of new issues which depend on position, and cultural/religious environment. These include:

- Adherence to the code of conduct prescribed for the profession by a related agency
- Regard for cultural values, traditions and practices in the work place
- Nurturing of harmonius work culture among coworkers, monitoring and assessment, awards and rewards,
- Mentoring for excellence and leadership roles,
- Adherence to the standards of safety, concern for safe practices at workplace, protection of life from risks, hazards, injury, stress, and illness,
- Protection of environment, adoption of sustainable and green engineering practices, proper maintenance and servicing,
- Concern for conflict of interest with one-self, employer, clients, or friends, cross-cultures
- Protection of intellectual property rights, transfer and commercialization of technologies, etc.
- Avoidance of favoritism and victimization in recruitments and promotions, grants, awards, peer reviews
- Elimination of feudal work culture, sycophancy, regionalism, & casteism
- De-bureaucratization of the management system with transparency and accountability
- Elimination of corruption for money and power

4.4 SOCIETAL ETHICAL CONCERNS

Engineering professional designs machines and develops products and processes for manufacturing using minimal energy and natural resources to minimize cost. No development is, however, possible without affecting our natural resources, ecology and bio-sphere. Dwindling energy resources, reaching limits of the carrying capacity of our ecosphere, and global warming are some of the warning signals to all global societies. Science is ethics-neutral but scientist-engineer and technology are not. Technology is a double-edged sword with useful and harmful sides. As a result, typical examples of such ethical concern in India are:

- Whether genetically modified (GM) plants, seeds, fruits and food should be allowed in India to serve and feed a billion people

- Whether CNG or unleaded petrol is a better fuel for our massive transport needs
- Whether aquaculture industry should be allowed in coastal areas
- Whether cement jetties harm the mangroves in Kutch
- Whether the height of the Narmada Dam is right
- Whether the Taj Mahal is at risk from air pollution
- Whether this or that polluting industry should be closed

With extraordinary recent developments in science and technology, in particular the evolving trio of Info-Nano-Bio has impacted ethical dilemmas literally like a "tsunami". Nanotechnology, Biotechnology, Bio-engineering, Biomedical, Clinical trials, Bio-agro-technology, Nano-agro-technology, GM products, Genetic testing, Biopiracy, Bio-weapons, Space technologies, Nuclear technologies, Information technologies, Communication technologies, Internet, World Wide Web (WWW), Weapons of Mass Destruction(WMD) are some examples. Even more serious ethical dilemmas are the rapidly emerging areas of stem cell research, genomics, proteomics, cloning (plants, animals, and, may be, humans), biopiracy, etc. The extraordinary power, authority and control of the World Wide Net, with its faceless communications, invasion of liberty and personal space, monitoring of workplace, controlled terrorism, etc pose serious global ethical dilemmas. These dilemmas have given birth to a new classification of ethics such as Bio-ethics, Medical ethics, Nanotech-ethics, IT-ethics, Space-ethics, etc which are of very serious global concern today.

4.5 ETHICAL RIGHTS

Those responsible for ensuring adherence to ethical values must be supported by legal, and moral, social and financial rights and also due recognition by government, employers, professional societies, colleagues and society. Some such rights are:

1. Right to bring any unethical practice to the attention of public (whistle blower)
2. Right to be protected legally, financially, professionally and physically if and when required
3. Right to participate in the activities of professional associations/ institutes/ academies
4. Right to protect confidentiality obligations and agreements
5. Right to protect IPR
6. Right to refuse an unethical assignment or job
7. Right to gender and minority equality

Those who are the victims of an unethical practice also must have appropriate rights. The **UN Declaration on Bioethics & Human Rights: Principles** (adopted by acclamation on 19 October 2005 by the 33rd session of the General Conference of UNESCO) is a relevant document in this connection which lists human rights as:

1. Human dignity and human rights
2. Benefit and harm (knowledge thereof)
3. Autonomy and individual responsibility
4. Consent: Prior, free and informed
5. Special protection to those who lack capacity for consent
6. Respect for human vulnerability and personal integrity
7. Privacy and confidentiality
8. Equality, justice and equity
9. Non-discrimination and non-stigmatization
10. Harmony and respect for cultural and religous diversity and pluralism
11. Solidarity and cooperation
12. Social responsibility and health
13. Sharing of benefits
14. Protecting future generations
15. Protection of the environment, biosphere and biodiversity

4.6 NURTURING OF ETHICAL VALUES

- "The growing concern over the erosion of essential values and an increasing cynicism in society has brought to focus the need for readjustments in the curriculum in order to make education a forceful tool for the cultivation of social and moral values". (National Policy on Education,1986).

  Despite this lofty declaration by educationists, very little has been done by any educational system in the country for nurturing ethical values among students of science and engineering. Business ethics are, of course, compulsory subjects in Schools of Management. A few academic institutions have recently set up Centers for Value Education which have designed courses for motivating interested students.

- Society for Scientific Values (SSV), a unique non-government Society, was registered by a number of prominent scientists and engineers of the country in 1987 to promote integrity, objectivity, and ethical values in the pursuit of science and technology in India. SSV has held several national and institutional seminars for sensitizing and motivating students, faculty and R&D scientists on ethical issues. The Society has prepared a code of conduct for research scientists and also acts as a watchdog for unethical practices in the knowledge community of India.

- In sharp contrast to our indifferent attitude, exposing science and engineering students to ethical values is mandatory in all institutions in developed countries. It is generally recognized that ethics cannot be taught in a formal way but must be cultured and nurtured through experience, analysis, introspection, sense of responsibility, and sensitization. Therefore, suitable material for open discussion on such issues as problem solving, case studies, moral decision making dilemmas has been prepared by academics in various western countries. Modular courses, integrated with regular science & engineering courses, taught by reputed and experienced scientists/engineers are helping to nurture future ethicists in many western countries.

- National Science Foundation, USA has prepared various motivational packages on ethics for integration with regular teaching-learning programmes for science and engineering students and which are recommended for delivery by prominent academics or professionals in an informal manner.

4.7 PROMOTION OF ETHICAL PRACTICES

Academic institutions and professional /societies/institutes/academies have a responsibility for developing and promoting knowledge-based professional ethics through awareness and motivational workshops, incentives and disincentives. With a few exceptions, most academic institutions in India do not have any defined code of conduct except for an unwritten honour code for students and faculty.

Compliance of a code of conduct is just as important as the code itself. There is certainly a need for a regulatory mechanism to monitor and to ensure compliance of the code. It is recognized that "no regulatory mechanism, however strong, can provide for the consequences of human greed, folly or corruption" (Tim Yeo). The reluctance of authorities to acknowledge the occurrence of proven cases of unethical practices and scientific misconduct even in prestigious academic institutions calls for a regulatory mechanism with some quasi-judicial powers. Some professional bodies which are seriously involved in designing codes of conduct are:

- The Institution of Engineers (India) was created in 1920 to promote and advance the art, science and practice of engineering and technology in India. It prescribed a set of Professional Conduct Rules in 1944 which have been revised and published as 'Code of Ethics" given in Appendix. However, no serious effort has been made to popularize the code. Nor is it a requirement for professional engineers to commit to adherence to the code.

- An Engineering Council of India (ECI) has been registered in 2002 to establish a confederation of engineering societies/associations. One of its objectives is to evolve codes of professional conduct and ethics for various engineering disciplines.

- It is noteworthy that Tamil Nadu is the only state in India which has published a code of conduct for civil engineering profession.

- Department of Biotechnology (DBT) and Indian Council of Medical Research (ICMR) and ICHE have prepared Ethical Guidelines on some aspects of Biotech, Biomedical, and Social Science Research on human health.

- Council of Scientific and Industrial Research (CSIR) has adopted the code of conduct of the Max Planck Society, Germany although it is not known to any CSIR laboratory.

- A detailed code of ethics, primarily applicable to scientific values and academic research integrity issues, has been developed by the Indian Academy of Science (IAS), Bangalore and is reproduced in the Appendix II. The document lists ethical guidelines and procedures for research. It also recommends a national regulatory authority for dealing with cases scientific misconduct.

- A pro-active Office of Research Integrity for life sciences set up by President Clinton in USA is an example to follow. Presently, formal and informal exposure to S&T ethics is mandatory for all students in most western academic institutions.

- Science, engineering and technology based Professional Societies in the west have evolved detailed codes of conduct and, in some cases, have set up Boards of Ethical Review as regulatory bodies.

- Many professional societies in USA have individually published their own canons or codes of ethics. In the US, the code adopted by the Engineers' Council for Professional Development (ECPD) is the most widely accepted. Some of the fundamental principles of these codes are :

The Engineer, to uphold and advance the honour and dignity of the engineering profession and in keeping with high standards of ethical conduct:

1. Will be honest and impartial, and will serve with devotion his employer, his clients, and the public.

2. Will strive to increase the competence and prestige of the engineering profession.

3. Will use his knowledge and skill for the advancement of human welfare.

There are more detailed statements under (a) Relations with the public, (b) Relations with employers and clients, and (c) Relations with engineers.

- A new umbrella organization has been formed in the U. S called the American Association of Engineering Societies (AAES) in 1980, which is seeking to establish the AAES Model Guide for Professional Conduct (AAES 1984) as a profession-wide code of ethics. The significance of the AAES with respect to social responsibility issues, however, lies in the relationship between the engineering profession and the public.

- The Federation of All European Academies (48) has advised the member academies to set up National Committees for Scientific Integrity and to evolve a code of conduct.

- The International Committee of Scientific Unions (ICSU) has set up a standing committee on responsibility and ethics in science and a very comprehensive report is available on its website. Further, ICSU has advised member academies to formulate country specific documents on Ethics in science and technology. In India, an inter-academy committee chaired by Prof. M.G.K. Menon is working on the document.

- UNESCO has evolved "Universal Declaration on the Human Genome and Human Rights". It has undertaken a variety of activities through its Division of Ethics of Science and Technology, International Bioethics Committee and the Division for mapping global Ethics Experiences.

  A Global Ethics Observatory (GEOB) established on Dec 15, 2006 has compiled three Data Banks on: Who's Who in Ethics; Ethics Institutions; Ethics Training Programmes and also a Division for mapping global Ethics Experiences. Further, UNESCO has set up two Ethics Chair Professorships in Israel and Russia and supports international seminars on various aspects of ethics in S&T in collaboration with member countries.

4.8 RECOMMENDATIONS

1. Ethical values must be recognized as being central to globalised knowledge-based economies of nations. Therefore, culturing, nurturing and sensitizing of ethical values among the knowledge communities of students, teachers and professionals must be accepted as an essential component of teaching-learning and sensitization process in all academic institutions and professional bodies.

2. Knowledge of professional codes of ethics and importance of professional morality should be integrated into the professional education system so that these are cultivated as a part of professional orientation of our students.

3. The privilege of the profession demands unconditional adherence to defined codes of practice and to its moral and ethical values. All academic institutions must have in-house monitoring and regulatory mechanism for ethical values through suitable "ethics vigilance" or "knowledge integrity" committees.

4. There is a dire need for a national, non-government, and quasi-judicial committee/commission to provide a think tank and a watchdog to deal with cases of unethical professional misconduct by individuals or corporations of national importance.

5.   The privilege of the profession to serve and excel demands unconditional adherence to defined codes of practice and to its moral and ethical values. This can best be protected by carefully developing codes of ethics and conduct for the professionals. A professional engineer should be required to make a written commitment to adhere to such a code. The code should be binding on its members and any deviation from the defined path should be strictly punishable or else the "privilege to serve" will become a "privilege to exploit".

6.   There is already a demand by such organizations as WTO to evolve globally acceptable and enforceable codes of professional ethics, morality and code of conduct for professionals and corporations engaged in knowledge-based global engineering services, manufacturing and trade. Such a code needs to be certified by a suitable international standards organization (ISO).

7.   Professional responsibility to adhere to ethical values must be supported by legal and moral rights, with suitable limitations and recognition by others of such rights.

8.   All institutions need to provide an empowered mechanism for respecting and protecting the whistle blower who exposes misconduct or unethical practices in an organization. A whistle blower's act needs to be adopted nationally.

APPENDIX I
INSTITUTION OF ENGINEERS (INDIA)
CODE OF ETHICS FOR CORPORATE MEMBERS

1.0 INTRODUCTION

1.1   Engineers serve all members of the community in enhancing the welfare, health and safety by a creative process utilising the engineers' knowledge, expertise and experience.

1.2   Pursuant to the avowed objectives of The Institution of Engineers (India) as enshrined in the presents of the Royal Charter granted to the Institution, the Council of the Institution prescribed a set of "Professional Conduct Rules" in the year 1944 replacing the same with the "Code of Ethics for Corporate Members" in the year 1954 which was revised in the year 1997.

1.3   In view of globalisation, concern for the environment and the concept of sustainable development, it has been felt that the prevailing "Code of Ethics for Corporate Members" needs review and revision in letter and spirit. The engineering organisations world over have updated their Code of Ethics.

1.4 The Council of the Institution vested with the authority in terms of the Present 2(j) of the Royal Charter adopted at its 626th meeting held on 21.12.2003 at Lucknow the "Code of Ethics for Corporate Members" as provided hereinafter.

1.5 The Code of Ethics is based on broad principles of truth, honesty, justice, trustworthiness, respect and safeguard of human life and welfare, competence and accountability which constitute the moral values every Corporate Member of the Institution must recognize, uphold and abide by.

1.6 This "Code of Ethics for Corporate Members" shall be in force till the same is revised by a decision of the Council of the Institution.

CODE OF ETHICS FOR CORPORATE MEMBERS

1.0 PREAMBLE

1.1 The Corporate Members of The Institution of Engineers (India) are committed to promote and practice the profession of engineering for the common good of the community bearing in mind the following concerns:

1.1.1 Concern for ethical standard;

1.1.2 Concern for social justice, social order and human rights;

1.1.3 Concern for protection of the environment;

1.1.4 Concern for sustainable development;

1.1.5 Public safety and tranquility.

2.0 THE TENETS OF THE CODE OF ETHICS

2.1 A Corporate Member shall utilise his knowledge and expertise for the welfare, health and safety of the community without any discrimination for sectional or private interests.

2.2 A Corporate Member shall maintain the honour, integrity and dignity in all his professional actions to be worthy of the trust of the community and the profession.

2.3 A Corporate Member shall act only in the domains of his competence and with diligence, care, sincerity and honesty.

2.4 A Corporate Member shall apply his knowledge and expertise in the interest of his employer or the clients for whom he shall work without compromising with other obligations to these Tenets.

2.5 A Corporate Member shall not falsify or misrepresent his own or his associates' qualifications, experience, etc.

2.6 A Corporate Member, wherever necessary and relevant, shall take all reasonable steps to inform himself, his employer or clients, of the environmental, economic, social and other possible consequences, which may arise out of his actions.

2.7 A Corporate Member shall maintain utmost honesty and fairness in making statements or giving witness and shall do so on the basis of adequate knowledge.

2.8 A Corporate Member shall not directly or indirectly injure the professional reputation of another member.

2.9 A Corporate Member shall reject any kind of offer that may involve unfair practice or may cause avoidable damage to the ecosystem.

2.10 A Corporate Member shall be concerned about and shall act in the best of his abilities for maintenance of sustainability of the process of development.

2.11 A Corporate Member shall not act in any manner which may injure the reputation of the Institution or which may cause any damage to the Institution financially or otherwise.

3.0 GENERAL GUIDANCE

The Tenets of the Code of Ethics are based on the recognition that –

3.1 A common tie exists among the humanity and that The Institution of Engineers (India) derives its value from the people, so that the actions of its Corporate Members should indicate the member's highest regard for equality of opportunity, social justice and fairness;

3.2 The Corporate Members of the Institution hold a privileged position in the community so as to make it a necessity for their not using the position for personal and sectional interests.

4.0 AND, AS SUCH, A CORPORATE MEMBER –

4.1 should keep his employer or client fully informed on all matters in respect of his assignment which are likely to lead to a conflict of interest or when, in his judgement, a project will not be viable on the basis of commercial, technical, environmental or any other risks;

4.2 should maintain confidentiality of any information with utmost sincerity unless expressly permitted to disclose such information or unless such permission, if withheld, may adversely affect the welfare, health and safety of the community;

4.3 should neither solicit nor accept financial or other considerations from anyone related to a project or assignment of which he is in the charge;

4.4 should neither pay nor offer direct or indirect inducements to secure work;

4.5 should compete on the basis of merit alone;

4.6 should refrain from inducing a client to breach a contract entered into with another duly appointed engineer;

4.7   should, if asked by the employer or a client to review the work of another person or organization, discuss the review with the other person or organization to arrive at a balanced opinion;

4.8   should make statements or give evidence before a tribunal or a court of law in an objective and accurate manner and express any opinion on the basis of adequate knowledge and competence; and

4.9   should reveal the existence of any interest – pecuniary or otherwise – which may affect the judgement while giving an evidence or making a statement.

5.0   Any decision of the Council as per provisions of the relevant Bye-Laws of the Institution

APPENDIX II

INDIAN ACADEMY OF SCIENCES

Scientific Values: Ethical Guidelines and Procedures
Bangalore 560 080

December 2005

1. PREAMBLE

Most professional bodies such as societies, associations, and academies expect their members to follow the highest ethical values in the conduct of their professional work. These values are based on universal moral principles like honesty, truthfulness, and fairness. For each profession, these values get translated into a separate 'code of conduct' statement. Such codes provide guidance on expected behaviour from the members under different circumstances and enhance the credibility of the profession in the perception of the public. They also tend to define an ideal which each member could strive to attain. The same ethical requirements apply to the scientific profession also. Science has many applied branches like engineering, and medicine, each having its own professional society and its own detailed code of conduct. These codes are based on the same basic moral principles but differ in detail because of the different activities of various societies.

The Indian Academy of Sciences has decided to concentrate primarily on those aspects which impinge directly on the activities normally pursued by its Fellows. Thus the major areas covered in this document include research, development, training, science management and policymaking. The activities pursued by the Fellows not only involve conducting research and publishing its results, but also interacting with students and trainees, colleagues and collaborators, members associated with science management and policy making, as well as with the press and the public. They also involve peer review, editorial work, as well as activities associated with development and management of technology. Each of these interactions has serious ethical implications.

The Fellows of the Academy have to remain sensitive to them while dealing with various situations encountered in the course of their professional work. This document has also paid special attention to the ethics involved in committee work in which many Fellows of the Academy participate. Science administration, management and policy-making too have received adequate attention. In all these areas, compromise on ethical behaviour can have a highly deleterious influence on the general atmosphere of the practice of science, and can result in a negative perception of the scientific profession.

Various areas of concern to the Fellows of the Academy are; (i) conduct of research; (ii) publications; (iii) training of students; (iv) interaction with the public; (v) science management; and (vi) ethics in technology-related issues. All these have been briefly discussed below. Mention has also been made of the positive role played by "whistle-blowers".

2. CONDUCT OF RESEARCH

The following activities pursued by scientists can be organized under this topic.

2.1 Data Collection

While conducting research, whether independently or jointly, it is necessary to ensure that data collected are reliable, properly recorded, and stored (even raw data) and there is absolutely no attempt made for either fudging data, or recording false data. The procedure followed should be described in enough detail that the results permit independent verification. Wrong data when reported in literature can cause confusion and in the short run may even prevent new ideas from being proposed.

Generating, recording, and reporting false data are fraudulent practices and need to be discouraged in all possible ways.

2.2 Sharing of Facilities

In most institutions, expensive items of equipment are few and have to be shared with other colleagues. Unfortunately sometimes this does not happen even when the equipment has been procured on the express promise that it would be used jointly. There is a consistent complaint of denial of adequate.

Access to equipment, particularly from junior colleagues as the person in charge makes disproportionate use of it for his/her own personal research. Fairness should be shown by following transparent procedures for time allotment.

2.3 Experiments Involving Human Beings or Animals

Guidelines and protocols announced by the various national agencies should be scrupulously followed. See for example, the announcements on "ethical guidelines for biomedical research on human objects, use of laboratory animals in research and training", "transfer of human biological material for research and development, stem cell research and therapy, and others which are available on the websites of Department of Biotechnology (New Delhi), Indian Council of Medical Research (New Delhi), the Ministry of Health, Government of India, the Indian Council of Agricultural Research (New Delhi) and the Medical Council of India.

3. PUBLICATIONS

Associated with publications, there are a number of issues like dishonesty in reporting, credit sharing with colleagues and students etc. where ethical behaviour is very important. Some of the important issues are mentioned below:

3.1 Plagiarism

Appropriating the already published results of others without proper reference is obviously dishonest. When exposed, plagiarism generally receives the highest publicity and the authors concerned and the system they belong to are put under tremendous pressure. In most cases, the concerned authors offer some explanation in their defense. However, sometimes they disown responsibility and even the knowledge of the papers' existence, claiming that the co-authors included their names without consulting them. Such disclaimers should not be accepted at face value, but should be looked into in more detail. Nobody should communicate a joint paper without the knowledge of the other authors.

There is a strong need to take punitive actions to discourage plagiarism. There is a general impression among the scientific community in India that those who indulge in this form of dishonest behaviour do not receive appropriate punishment, and escape relatively unscathed. Stronger and more consistent action would redress this situation.

3.2 Duplicate Publications

Sometimes an author publishes the same article at more than one place without mentioning that it has appeared before in the same or similar form. This practice is against professional ethics and must be avoided.

3.3 Order of Authors

Some societies specifically suggest the order in which the names of authors should appear. However, the basic requirement from the ethical point of view is that each author should receive adequate credit.

A transparent procedure of deciding the order should be formulated and followed scrupulously.

3.4 Undeserved Authorship

Some scientists having administrative control over others permit or require their names to be included in publications and patents in which they have made no scientific contribution. This practice is not acceptable, as it is very much against the spirit of doing research and reporting results.

3.5 Purity of Data

The data reported in a publication should be authentic and should not be based on biased observations or fabricated. It is necessary that sound data-collection practices be followed. Even if only part of the data is being reported, the complete data and computer programs, in their original form, should be retained.

3.6 Sharing Responsibility among Authors

Increasingly, papers are being published where expertise of many scientists has been pooled together. So much so that each one may have only limited knowledge of what has appeared in the paper. Thus in the case of misconduct regarding one of the sections, it may not be appropriate to hold all the authors equally responsible. It is recommended that the authors generate a written statement on the responsibility of each author. A copy of this document may be kept with each of the authors and used in case the need arises for fixing responsibility. In the absence of such an understanding, all authors should own equal responsibility for the whole publication.

3.7 Peer Review

Many scientists act as reviewers for manuscripts submitted for publication as well as project proposals submitted for financial support. In both cases, they get access to information and ideas which have not yet been published. It is important for them to ensure that this advance access to information is not unethically exploited by them for their own benefit.

4. INTERACTION WITH THE PUBLIC

Scientists must ensure that statements made in public are dependable and balanced. They should avoid making exaggerated claims to the press or at public meetings. Scientists should be held accountable for the claims they make.

5. TRAINING PROGRAMMES

Most scientists are teaching as well as supervising research and are hence deeply involved in training programmes. The training is likely to inculcate higher ethical values in students if the institution concerned provides an enabling environment where honesty, truthfulness and fair play are practiced as a matter of routine. Firstly, the training should be such that the student is continuously exposed to a high level of ethical behaviour. Secondly, special training should be imparted to them about contemporary ethical values. The major issues concerning the ethical environment are associated with teaching, assessing, distribution of students to various supervisors and the behaviour of potential role models.

5.1 Teaching and Assessment

Teaching should be taken very seriously, with considerable effort being put in to communicate effectively with students. Above all, there should be easy access for the students to the teacher to clarify any issues. Student assessment should also be transparent and fair without any bias.

5.2 Student Allotment and Research Supervision

While allotting students to various faculty members, the interests of both the students and faculty members have to be kept in mind. Different institutions may handle this differently, but to maintain fairness, the procedure for allotment should be decided and announced beforehand and should have the general approval of the concerned members. The procedure should be followed in a transparent fashion.

5.3 Direct Ethical Training to Students

The Academy emphasizes the need for direct training of students as a way to generate an ethical atmosphere in research institutions. A number of methods may be employed, including special formal courses, discussions and workshops, formation of local ethical committees, and generating articles on ethical issues. Independent of their formal training, students tend to emulate their role models and their supervisors. If they see their role models indulging in unethical practices, the effect of formal teaching is likely to get diluted. Therefore direct ethical training needs to be supplemented by internal discussion and monitoring of practices.

6. SCIENCE MANAGEMENT

A lack of ethics in science management can have very far-reaching deleterious effects on both research and education. As many of the decisions in this general activity are taken through committees, it is here that the highest standards of fairness and balance are required. Some of the areas which impact Indian science, and where committees make the basic recommendations, are concerned with recruitment, assessments and promotions, project grants, performance awards and science policies. Each one of them is briefly discussed below.

6.1 Recruitment and Assessment

The quality of faculty decides the standard of an institution to a very large extent. Once low quality recruitment has been made, the effect can last for many decades. It is unethical to discard candidates of higher quality and select others based on extraneous considerations. It is equally unethical to deliberately permit personal prejudices and biases to dominate during the process of selection. It may be noted that a selection procedure can be unethical even though not illegal, and therefore it is all the more important to remain sensitive to these issues while participating in the decision making process. The selection of committee members who are known for fairness and balance rather than pliability is important.

It is unethical to discriminate, in recruitment of scientists (both students and faculty) on grounds of gender. Committees should make every effort to ensure that such biases do not enter into their decisions, and should be cognizant of the National Science Policy (Department of Science and Technology, New Delhi, 2003, www.dst.gov.in/st_policy.htm) in this regard. There should also be no bias against hiring spouses in the same department or institution. All the ethical guidelines mentioned above for recruitment are equally applicable for promotions.

6.2 Project Grants

Most active scientists seek financial support for their research. The usual route is through submission, assessment and grant of projects. These projects are normally first peer-reviewed and then discussed by a committee. Sometimes even more than one committee examines the proposal before granting funds.

As both peer review and committee discussions involve a certain degree of subjectivity, it is very necessary that the highest ethical standards are observed by committee members as well as referees.

Decisions on grants based on biases of any kind are highly unethical.

6.3 Awards

In recognition of excellent performance there are awards instituted by the Government as well as by private bodies. To select one or two persons from a field of close competitors is indeed hard, and therefore it is important for the selection committee members to study each case carefully. To maintain the sanctity of these awards, unbiased decisions must be made.

6.4 Policy Issues

Some decisions taken by the Government are based on inputs provided by scientists. Typical examples may be GM crops, stem cell research, human cloning and other emerging technologies.

It is important that scientists provide honest and well thought-out views rather than giving in to commercial, social or political pressures. It is only when scientists maintain the highest ethical standards following the dictates of their own conscience, in all committees, that the credibility of the scientific community will be at the highest level.

7. ROLE OF WHISTLE-BLOWERS

Some times individuals are forced to expose irregularities when the system remains indifferent or actively suppresses the misconduct reported by them. Such whistle-blowing needs careful investigation and follow up action. Whistle-blowers perform an ethical public function at risk to themselves and deserve not only protection but also our admiration.

8. ETHICS IN TECHNOLOGY-RELATED ISSUES

There are a number of Academy Fellows who are engaged in applied research, involving technology development and commercialization. These areas have their own characteristic ethical issues, having to do with sustainable development, technology acquisition, sale and transfer of technology, sharing of intellectual property rights, industrial safety, and other matters like environmental loading. Existing ethical guidelines in specific areas must be identified and followed. One should also be sensitive to areas like dual-use technologies for which ethical guidelines are still being debated.

9. REGULATORY MECHANISM

It is recognized that some incidents involving unethical behaviour are best handled locally. However, many types of misconduct like plagiarism are of special importance as they have an adverse effect on the credibility of the entire scientific community. Ideally, there should be a centralized scientific body to handle all issues pertaining to scientific ethics.

In the absence of such a national body, the Academy has generated a regulatory mechanism, applicable to the conduct of its Fellows, and hopes that other institutions will follow similar procedures until a centralized body becomes functional. The final decision about the reported misconduct of a Fellow will be taken by the Council of the Academy, based on inputs provided by the Panel on Scientific Values constituted by the President of the Academy.

The detailed procedure to be followed is given below:

A. (i) Any complaint against a Fellow of the Academy should be addressed to the President of the Academy.

  (ii) The complaint should be signed by the complainant and should have his/her complete address.

  (iii) The complaint should be accompanied by appropriate documents authenticating the complaint.

  (iv) The complainant should not publicize the matter until a final decision is taken by the Academy.

B. The President will decide about the suitability of the complaint for further processing.

C. If cause is found for further investigation, the President will obtain a declaration from the complainant that he/she takes full responsibility for the genuineness of the complaint and is prepared for consequences, such as exposure, in case the complaint is eventually found to have been filed with malafide intentions.

D. The President will then pass on all the documents, after suppressing the complainant's identity to the Panel on Scientific Values of the Academy for a complete investigation.

E. The Panel will contact the Fellow against whom allegations have been received and obtain his/her version.

F. Should the Panel feel necessary, it can have personal discussions with the Fellow concerned.

G. The Panel may invite any expert, who can help in the investigation through his/her specialized knowledge.

H. The Panel, on completion of its work, will submit a report to the Council. This should be done within three months of receiving the documents from the President.

In case the Panel finds the allegations to be substantiated, it can suggest possible penalties in its report to the Council.

I. The Council will decide on the actual action to be taken against the Fellow.

J. If the Panel concludes that the allegation was made with malafide intention, the President may disclose the name of the complainant to the Panel, which in turn will obtain the person's explanation in writing and also in person, if necessary.

K. In this case too, the Panel will submit its report to the Council containing appropriate recommendations.

REFERENCES

1. C.G. Krishnadas Nair "Engineering Ethics"Harishree Publishing Co, Bangalore, 2006.

2. A.N. Tripathi "Human Values", New Age Int Publishers, 2008.

3. S. K. Chakraborty "Ethics in Management: Vedantic Perspectives", Oxford University Press, Delhi, 1995.

4. A.N. Tripathi "Human Values in the Engineering Profession", published by IIM, Calcutta,1997.

5. IEEE US Publication "Guide to Ethics", IEEE catalog no. UHO, 1982.

6. S.K. Chakraborty "Value Ethics for Organisations" Oxford University Press, New Delhi, 1998.

7. Caroline Whitbeck, "Ethics in Engineering Practice and Research", Cambridge University Press, 1998.

5

Faculty Related Issues

"There is nothing training can not do. Nothing is above its reach. It can turn bad morals to good; it can destroy bad principles and recreate good ones; it can lift men to angelship."

— Mark Twain

5.1 FACULTY SHORTAGE AND FAILURE TO ATTRACT AND RETAIN AN ADEQUATE NUMBER OF TALENTED PROFESSIONALS FOR A TEACHING CAREER

The technological education system in India is facing an acute shortage of teaching staff in most engineering colleges and universities. With a pyramid of institutions starting with IITs at the apex, the system has to deal with the NITs (National Institutes of Technology), GECs (Government Engineering Colleges), and PECs (Private Engineering Colleges) for the Baccalaureate programs. The question of quality is the most important issue. The quality of faculty, staff and students constitute the core competence of any educational institution. In India, the quality of faculty is a matter of utmost concern. Since the size of the student body is very large, the need for faculty is also high. At present, it is estimated that about 120,000 faculty members are needed for various engineering colleges. The educational qualification for these teachers is desired to be at the doctoral level. However, such a trained manpower is just not possible in next five years. Most of the institutions are not able to appoint faculty members even with Master's degree. The perception of the society is that the compensation package in academia is not attractive at all. Many talented individuals do not prefer to pursue an academic career. As a result we are concerned about adequacy of the numbers of trained faculty for engineering education. Besides the professional degree at a suitable level, the faculty members should have some knowledge of teaching pedagogy. Unfortunately, the knowledge of teaching pedagogy is poor even at the elite institutions. This results in a teaching community that cannot connect themselves effectively with the students. In short, the faculty members in many cases are inappropriately positioned.

5.2 UPGRADATION OF QUALIFICATIONS OF EXISTING FACULTY

The AICTE (All India Council for Technical Education) recognized the issues of quality improvement several years back. Many schemes have been implemented by AICTE so as to address the issues of faculty development. The QIP (quality improvement program) is a very successful scheme. Several teachers from various engineering colleges are selected and deputed at different IITs / IISc and Institutes of repute to complete their MTech or PhD programs. Once they finish their academic work, they return to their parent institution and provide mandatory service for a certain number of years. This scheme is a very successful scheme. Since the size of the system has grown enormously, the scheme is not able to take on the load as needed at present.

The Early Faculty Induction Program (EFIP) was started by AICTE in the recent years. However, this scheme has not been successful. Young students upon graduation would like to keep their options open and would not like to commit to a career during their student days. The lure of software industry as well as the prospects of high salaries for those jobs did not help in achieving a high degree of success for EFIP scheme.

The short term courses of AICTE as well as ISTE were very popular at one time. These courses of one or two week duration have been very effective to expose the teachers of engineering colleges to new areas of science and technology. Once again, these courses are now not so popular for several reasons. The budget approved is found to be very much inadequate. Some of the topics covered in these courses may be of interest to a limited number of researchers but not to a large number of teachers. The work of developing course material is very demanding. In some cases, teachers are not given leave due to the demand of the academic calendar in their parent institution.

As mentioned earlier, the Quality Improvement Programme (QIP) of AICTE is one of most successful educational initiatives in the country. It has brought about a sea change in the academic ambience of a large number of NITs and Government Engineering Colleges. A significant number of teachers in various engineering colleges have successfully made career advancement in the fields of science, technology, and management through QIP. The scope of the QIP has to be broadened so that the entire population of young teachers serving in various Government and private engineering colleges are benefited from the programme.

5.2.1 Revision of Existing Norms

The government is keen in extending the domain of activities of QIP. It is felt that the teachers in the private engineering colleges have to be given opportunity to advance their career through QIP. Currently about 100 PhD positions and close to 400 MTech/ME slots per year are supported by the QIP. This can straightway be extended to 1000 PhD positions and about 2000 MTech/ME slots. The current scholarship, contingency etc. have to be revised. Typically the modifications could be as the following:

Category	First year	Second Year	Third year
MTech/ ME	Scholarship: at par with the MTech students at the IIT system. Contingency: Rs 15000 per year.	Scholarship: at par with the MTech students at the IIT system. Contingency: Rs. 15000 per year.	Not Applicable
PhD	Scholarship: at par with the PhD students at the IIT system. Contingency: Rs. 15000 per year.	Scholarship: at par with the PhD students at the IIT system. Contingency: Rs 20000 per year.	Scholarship: at par with the PhD students at the IIT system. Contingency: Rs. 25000 per year.

Usually, the teachers from the Government Engineering Colleges obtain sponsorship (full salary) from the parent institutions while working as a QIP scholar. The sponsorship is based on the condition that the teachers go back to the parent institutions on completion of the PhD and serve for a period of minimum 3 years. The teachers from the private engineering colleges usually do not enjoy this benefit. Therefore securing a sponsorship from the parent institute should cease to be the mandatory requirement for participation in the QIP programme. However, it will be a firm binding to each PhD scholar under the QIP scheme that he/she has to serve an academic institute of the country for a period of at least five years on completion of his/her PhD. For the QIP scholars at the master's level this binding will be there for three years. After completion of PhD, it should be mandatory for the QIP scholar to guide at least one student for PhD at the institution where he is employed. A detailed plan for this has been outlined in Section 5.

5.2.2 Adjunct Faculty for Sharing Expertise with Industry

The academic institutes should offer research professionals in national laboratories opportunity to serve as the faculty members over a specified duration and to interact with faculty in research. Such interactions are possible through introducing visiting professorships or adjunct professorships. Some of the potential benefits are: academic departments can obtain the expertise and services of research professionals or staff without having to make permanent commitments for employment. National laboratories and academic departments can share in the expensive equipment or research facilities. There are other major reasons for better interaction as well. More interaction between academic departments and the research professionals could provide the balance between the complementary components.

5.2.3 Some Strategic Policies to be Adopted by the Government

The government has to make postgraduate qualification mandatory for recruitment to many positions in the engineering profession. Also it should be a requirement that the lecturers with BTech should acquire MTech within five years and a PhD within a specified time. The PhD degree must be a desirable qualification for teaching at postgraduate level. Government should encourage industry to support postgraduate programs or to establish scientific infrastructure with tax rebates. New and emerging areas have been identified which include Biotechnology, Synthetic Biology, Environmental Engineering, Energy Management, Remote Sensing, Sustainable Development Technology, Materials Technology, Nanotechnology, Polymer Technology, Food Process Engineering and many other fields of direct relevance to Industry. The usual belief that our country has a large pool of scientific and engineering workforce is no longer valid especially in relation to doctorates in engineering. It is also to be understood that companies need good quality PhD level researchers to be able to compete in the world market with quality products. The important aspects of "periodic upgrading" or "life long learning" have not been so far adequately addressed. The impact of the IT revolution on conventional areas of engineering has not been analyzed critically. An appropriate mechanism of partnership between the Ministries such as Power, Petroleum, Coal, Railway, Steel and Mines and the MHRD is needed for encouraging engineering education. Accomplished professionals from the organizations run by the above ministries may be encouraged to spend a semester in the academic institutions and engage themselves imparting their professional knowledge to the students.

5.3 INSTITUTIONALIZING FACULTY DEVELOPMENT MECHANISM

During the past ten years or so it has been seen that the private colleges are unwilling to release their faculty for further studies. The preoccupation of the faculty members with the system is continued without adding new input to the existing knowledge. Following is the solution of this heretofore unattended problem. Many of these faculty members may be encouraged to attend advanced courses using (i) a repository of Video courses created by very well known experts in the field (ii) a website that hosts the web-based learning material and (iii) the live lectures delivered in distance learning mode based on satellite and internet technologies. Recently proposed Virtual Technical University (VIRTUAL TECH) will play a major role in this initiative.

It is expected that VIRTUAL TECH will develop course material using the expertise available in the country. The course material will be in the form of video tapes, web based learning information as well as course ware in the form of reference material. As mentioned earlier, the students at VIRTUAL TECH are primarily teachers at various Institutes or Universities, who are unable to obtain study leave as full time students. They may like to attend the courses of their choice and earn credit for the courses that are close their areas of interest. A course of 1 credit under this scheme would comprise of 10 hours of lectures (instructions), 2 hours of tutorials, 8 hours of home work and 1 hour of test. The tutorials would be used to assign home work, provide necessary help in sorting out difficulties in solving the problems. A course of 2 credits would have 20 hours of lecture, 4 hours of tutorial, 16 hours of home work and 2 hours of test. Likewise a course of 4 credits would have 40 hours of lecture, 8 hours of tutorial, 32 hours of home work and 4 hours of test.

It is proposed that VIRTUAL TECH will provide a credit pass book for every student. Every course will be entered in the book along with the grade obtained by the student. The students may wish to conclude a program with the accumulation of some credits. Depending on the level of credits accumulated by a student, he or she will be awarded a certificate in an area or a diploma. In case, a student has completed sufficient number of credits and also has carried out a project, he or she will be awarded a post-graduate degree. In this way, VIRTUAL TECH will provide different exit routes for the students. The VIRTUAL TECH has to emphasize quality in addition to quantity. One without the other is not enough.

5.4 EXPANDING POSTGRADUATE EDUCATION AND RESEARCH MANY FOLD

Since VIRTUAL TECH may take time to come into operation, a scheme has to be identified that can be readily directed to the teachers at the private engineering colleges. The Video and Web-based lectures prepared by the IIT and IISc faculty for the NPTEL program, can be supplied to the private engineering colleges. The young teachers can become facilitators-cum-instructors for such lectures. Also a provision can be made for an open-house (once in a semester) where the teachers involved in using a NPTEL courseware will get opportunity to meet the content creator. They will be allowed to brainstorm over several aspects of the course in such workshops. The AICTE has to make it mandatory for the private engineering Colleges to use the NPTEL courseware in the disciplines of Computer Science, Electronics, Electrical, Civil and Mechanical Engineering. Also the available NPTEL courseware for the core courses must be used.

Another suggested scheme is as the following. Some selected faculty members of the private engineering colleges will be invited to IISc/IITs/Institutes that have QIP Centres (ref. HPC/MAK 6/8/2006 of Prof. M. Anandakrishnan) each year during the summer. The selection procedure will ensure that everyone interested gets a chance to participate in the programme. The invited faculty members will participate in the summer schools on the specific topics organized by the host institutes. The summer schools will cover the important subject areas related to Mechanical Sciences, Electrical Sciences, Mathematical Sciences, Physical Sciences, Chemical Sciences and Pedagogy. Each summer school will have rounded up teaching material on a specified theme, starting from basic underlying principles to advanced applications. These programmes will follow the philosophy of Sequential Summer Schools that were quite popular some twenty five years ago. The participating teachers will earn credit for attending such summer schools. One will earn five points for attending a summer school of duration of 10 days. Similarly, eight points may be allocated for attending a summer school of duration of a fortnight. These points will be counted while considering performance based promotion of a teacher. The candidates with more such points will get preference during the selection of QIP scholars.

Another idea is to introduce a novel postgraduate program. Such postgraduate programs may be started in some good private/ newly built government engineering colleges in partnership with the IITs. The purpose of such programs is to produce trained manpower with optimal costs. The IITs may be allowed to collaborate with the satellite campuses within the premises of such selected colleges to run off-campus post graduate programs in some chosen areas. The collaborating institutions will provide infrastructure facilities and residences for the students enrolled in these special postgraduate programs. Each IIT will be expected to develop five electronic classrooms with broadcast facility in their own campuses in order to transmit online or offline lectures using appropriate transmission facilities. The experiences of IIT Bombay, Delhi, Kanpur, Kharagpur and Madras on delivery of lectures through VSAT and multicast modes are to be made use of in accomplishing this task. The postgraduate degree will be awarded by the partner institutes/ universities, while the IITs will play the role of mentors. Each partner institution will house classrooms for IIT lectures and will have robust connectivity to the internet and intranet. The institution will identify two/ three young faculty members in each discipline who will be responsible for the campus administration and the academic administration of the courses. They will work closely with the service providers for managing the campus network and student services. They will function as the facilitators and primary interface with the partner IIT. Recommendations are that the colleges hire at least a few retired faculty members from IIT who are familiar with the teaching ethos of the IIT system. The primary role of such faculty members would be supplementing the teaching and answering queries of the students. The interactions among the facilitators, elderly faculty members and the students are expected to bring about satisfaction of all. Thus the virtual university will enrich the host institutions in the long run with better teaching resources and more knowledgeable faculty.

Since the government run systems alone are not able meet the skyrocketed demand for higher education, the expansion of private higher education can be blended with the public initiatives in order to meet both the need for increased capacity and the desire for innovative and more often than not employment-oriented, programmes. It seems that a lot more investment is needed in postgraduate level education. A joint government-private sector board may be established to steer innovative postgraduate courses at various locations in the country. They should evolve the financial means in support of such postgraduate education. The private sector has to understand the far reaching consequences of such initiative. They benefit directly from the educational product.

5.5 POST DOCTORAL WORK ON PROJECT MODE

Creativity comes from an inner innovative drive to accomplish novelty. Only when we seek excitement and do things for the sheer fun of it, do we crack a break-through. But the inner self-expression needs a proper atmosphere which unfortunately does not prevail in many Government/ private engineering colleges. The QIP scholars train themselves well and pick up ideas during their stay in the Institutes of higher learning. Often they become stagnant on their return to the parent Institutes because of lack of facilities and encouragement. It is suggested that each QIP scholar, who has successfully completed PhD, be given a project. The proposal submitted by the former scholar will undergo a scrutiny and it will be sanctioned, if found reasonable. The project will have the budget allocation for equipment/ computational facility (as appropriate), contingency and travel. In many situations, the equipment grant is crucial since the laboratories in most colleges are poorly endowed in terms of experimental facilities. More importantly, the project will support a PhD student. Basically there will be provision for a SRF with a scholarship of Rs.15000 per month for a period of 3 years extendable by one more year. The former scholar may like to involve his erstwhile thesis advisor in the project, if needed. The upper limit of funding for each project can be up to Rs. 25 lakhs. This will bring about a new research ambiance and the group of young enthusiastic faculty members will be able to produce spectacular results commensurate with the present day competitive environment.

5.6 CONCLUSION

Recent developments in the understanding of functioning of human brain and the learning styles indicate that key to quality education is to generate passion for learning. In order to generate passion for learning, AICTE and other Government Institutions have to adopt a proactive role. QIP is a very successful initiative. During the next phase, QIP has to take the responsibility of grooming the faculty members across the country. It has to review and redefine goals and targets for continuous improvement of qualified teachers in adequate numbers. In order to encourage innovation in teaching, teaching methodology and research, the issue of career advancement of the teachers at various engineering colleges is of paramount importance. The Quality Improvement Programme (QIP) has to enter in to a new paradigm. The Colleges have to understand that faculty with only a Master's education will not be able to bring the country to a high level of technical competence. The quality of technical education in the entire country will reach a level of excellence, if the moral of the teaching community remains enthusiastic, innovative and inquisitive. All these will be possible if more teachers get the opportunity to pursue their doctoral studies (PhD) at the well known Institutes of higher learning. The Government Institutions and AICTE have to share a fundamental focus — the development of intellectual human capital.

REFERENCES

1. National Policy on Education (1986/1992/2000) available at:
    www.aicte.ernet.in/National_Policy.doc
2. Pawan Agarwal "Higher Education in India: The need for change", 2006.
3. P.B. Sharma, Knowledge Power in Higher Technical Education, DCE Publication, 2008.
4. M. Anandakrishnan, HPC/MAK 6/8, 2006.

6

Undergraduate Curriculum and Evaluation Mechanisms

"Education is the manifestation of the perfection already in man."

— Swami Vivekananda

6.1 PREAMBLE

Engineering education is passing through a critical time in India. The general perception in public mind is that most of the employment opportunities in our country are in the area of Information Technology. This has resulted in reduced interest of students in hardware related disciplines. Attracting students to 'core engineering' is becoming difficult. At the same time, enormous changes in all disciplines of engineering and technology are taking place around the world, which are so vast and rapid that keeping pace with them is becoming increasingly difficult. An engineering education must include both hardware related as well as soft engineering knowledge and skills and also interpersonal skills. There is a far greater emphasis now to make the curriculum broad based and flexible to meet the greatly increased diversity of needs. Over and above, it should have a national perspective covering the economic and industrial growth.

In the early days of engineering education, the colleges/ universities trained students well for the then needs of industries and the government organizations dealing with irrigation, power, railways and other services to the society. During the first phase (1847-1950), the engineering education was know-how oriented and greatly dealt with practices. It fulfilled its objective well. In contrast to Mechanical, Electrical and Civil Engineering education, the education in the discipline of Electronics Engineering was more science based, which started around 1955 and most of the pioneers in this discipline came from Physics background.

The scenario of engineering education changed rapidly after the World War II.

Significant advancements in various branches of Science took place within 15 years. As a consequence, the Engineering and Technology made an enormous impact on the living conditions of the people via a variety of innovative consumer products. It became imperative for the engineering education to change its orientation and adapt the newer concepts of science. Continuous changes became a routine feature of engineering curricula. A change in emphasis from engineering practice to engineering science was implemented in all good institutions across the world. The developments in UG curriculum at the IITs during the early years were very unique in the Indian context.

The UG programmes of the IITs were oriented with the modern outlook and innovative spirit. More emphasis was put on basic sciences and mathematics. The courses in humanities and social sciences became important. IITs adopted in a more structured form the continuous system of evaluation of students, which was a distinctive feature of engineering education right from the beginning.

As mentioned earlier, today there are about 2300 Engineering Institutions in the country and the intake capacity of AICTE approved institutions of technical education alone is more than 8,00,000 undergraduate students. These institutions represent one of the largest educational systems in the world. The challenges posed by this rapidly evolving system in our country need to be understood. For effectively addressing the strategy of nurturing a large number of highly intelligent post-secondary students in a proper manner, quality of engineering education emerges as the issue of prime importance.

6.2 CURRICULUM

Engineering education is expected to produce trained manpower for maintaining and advancing technological growth in the country. The relevance and usefulness of engineering education have a strong bearing on the technological growth and the prospects of the country. The educational strategy should help to develop a knowledge economy. The systems involved in this endeavor should strive for furtherance of knowledge. It is needless to say that the engineering education should equip the people with new and relevant skills to embark on innovation with new ideas.

The Universities and Institutes have a major responsibility of analyzing and understanding the state of existing knowledge and its future perspectives, creating new knowledge, and assimilating the new knowledge that is being generated world over. The students have to understand the effect of globalization on the society. Many of them have to be prepared to become Entrepreneurs. *The Academic Programmes should have enough flexibility to cross the disciplinary boundaries, and to play with innovative concepts. The students should be encouraged for creative activities, be it in the realm of esoteric world, or in the area of hands-on experience.*

The future engineers will have to be more analytical and creative. Engineering practices based on empirical studies and data alone, will not be enough to give us cutting edge advantage. The engineers will be depending more on general analysis rather than relying on specific fault identification. They will have to be innovative; growth oriented and will have to respect entrepreneurship. The present generation has greater inclination for informal learning and is fond of flexible job market.

Indian engineers and technologists have done very well in Information Technology. Such successes are indeed needed in other areas, involving design and manufacturing. Miracles are needed in the areas of Power Generation and Distribution, Construction Technology, Water Purification, Biotechnology and environment protection.

The engineering graduates of advanced countries possess a high level of confidence in handling unknown problems. Many of our graduates devote their learning for the purpose of examinations. This deficiency has to be removed through setting up appropriate courses to make them innovative. They have to learn problem solving and design skills. The role of future engineers will be to integrate the concepts of economic prosperity with engineering knowledge and skills and apply their knowledge through innovations for the social and economic benefits of the country.

Taking an overall view, engineering education in the country must have a minimum but essential common structure and features. This will be necessary to assess the level of competence, which the education system enables the graduating students to have for their mobility and for the recognition of their degrees and diplomas. The common structure should prescribe the duration of the course of study, the minimum number of course credits and a flexible framework of distribution of the courses in terms of minimum requirements for humanities, science, core engineering subjects, electives for specialization and other requirements such as projects, foreign languages and value addition courses. These courses should also have coherence and reflect a unity of purpose. The current trend world over is to structure the academic programmes in a Credit based academic system. The central idea is that a student has to earn a prescribed minimum number of Credits in order to fulfill the requirements of graduation. The Credits are defined for the activities of the teaching-learning programme built into the curriculum.

Definition of Credits

The academic load and the credit for a given course unit is decided by the following **L-T-P-C** calculation.

where,   the number of one-hour lectures per week = **L**

the number of tutorial-hours per week = **T**

the number of practical (laboratory) hours per week = **P**

the Credit point for the course = **C**
One hour lecture in a week = 1 Credit
One hour tutorial in a week = 1 Credit
Two hours laboratory work in a week = 1 Credit
The Credit nomenclature of a course unit is denoted by **L-T-P-C**

For example, the Credit nomenclature for the first course in Mathematics may be identified as 3-1-0-**4**

6.2.1 The Science Component in Engineering Education

The excellence of the Baccalaureate programme at any world class Institute/ University is linked to its strong base of science education. Young students during their undergraduate education mature rapidly. Their personal and professional aspirations also change sometimes. The graduates must be able to apply their general educational experience and the knowledge of mathematics, physics, chemistry, biology and engineering to a wide variety of careers such as industry, higher studies, research, management, et al. Education in humanities and sciences expose the students to the work and thought processes of some of the best minds of the human civilizations. Moreover, due to their ubiquity, modern developments in science play a major role in the development of technology. Today biology is playing the most important role in many of the emerging technologies. A balanced course in biology is indispensable in the modern engineering education.

A minimum of 36 Credits of combination of Mathematics and Basic Sciences must be there in the Curriculum. Thus there should be three courses in Mathematics each course having 4 Credits (L-T-P-C: 3-1-0-**4**). Similarly two courses are needed in Physics (each 3-1-0-**4**) covering a total of 8 Credits. At least one course in Chemistry (3-1-0-**4**), one course in Biology (2-0-4-**4**) and one course in Earth and Environmental Sciences (3-1-0-**4**) would be the minimum requirement. The course should be a pointer towards "Green Technologies". One Laboratory course (1-0-6-**4**) would be needed to cover the basic experiments in Physics and Chemistry.

Mathematics includes elements of linear algebra, differential and integral calculus, differential equations, statistics, numerical analysis and discrete mathematics. The subjects in Basic Sciences are intended to impart an understanding of laws-of-nature and relationships through the use of analytical techniques.

6.2.2 Courses in Humanities

A minimum of 12 Credits in Humanities, Social Sciences, Arts, Management, Economics and Communication-skills should complement the Science and technology contents of the curriculum.

Some areas of study are considered to be essential in the education of an engineer. The social awareness and knowledge on economic developments are almost mandatory requirements. Among 12 Credits, eight (8) Credits are to be completed within first four semesters, whereas other four (4) Credits are to be completed during the remaining period of the curriculum. Among the eight (8) Credits that are to be covered during the first four semesters, one course should be on Communication Skills (2-2-0-4). Students are supposed to do a variety of technical writing during their stay in the Colleges/ Universities. Even otherwise, communication is the most important skill one should acquire. The students should improve their oral and written communication skills during the first two years of their training. **One dedicated course is essential for this purpose.** Provision must be made to develop capability of the individuals to communicate adequately, both orally and in writing.

6.2.3 Engineering Sciences (Compulsory), Practice Related Courses and Engineering Sciences (Options)

Engineering science subjects help the students to carry their knowledge further toward creative applications. They may involve the application of principles of mechanics, development of analytical/ numerical techniques, modeling, simulation and understanding experimental procedures. A minimum of 12 Credits is recommended as the Compulsory Engineering Science courses. These are Fundamentals of Computing (2-0-4-**4**), Introduction to Electronics (2-0-4-**4**) and Engineering Mechanics (3-1-0-**4**). Two courses on engineering practices are needed for acquiring engineering skills. These are Engineering Graphics (2-0-4-**4**), and Introduction to Manufacturing Practices (1-0-6-**4**). The Engineering Science (compulsory) and the Engineering Practice courses are to be taught across the disciplines. The Practice courses are to be modified with the perspective of enhancing employability.

Engineering Science courses can be further extended. An integrated study of engineering analysis, design, and engineering system for professional activities are planned and carried out to stimulate creative and imaginative thinking and making full use of the basic and engineering sciences.

Application to the identification and solution of practical engineering problem are to be stressed. These subjects may be categorized as the Engineering Science Options. Such courses include Thermodynamics, Solid Mechanics, Fluid Mechanics, Transport Phenomena, Electrical Technology, Elements of Materials Science, Geosciences, Data Structures, Environmental studies and other subjects pertinent to the disciplines. The Engineering Science Options may also include some subjects that impart an overall understanding of other engineering disciplines. There should be 16 credits for the Engineering Science Options.

Table 6.1 shows the courses appropriate for first four semesters of the B.Tech./ B.E./ BSc Engineering Program.

Table 6.1

Type of Courses	*Credits*
Mathematics and Science Courses	36
Communication Skills and Humanities	08
Engineering Science (Compulsory)	12
Engineering Practice Courses	08
Engineering Science (Options)	16

6.2.4 The Professional Courses

The Department specific courses will be taught in the subsequent four semesters. However, a significant emphasis on engineering design is needed. Engineering-design, integrates mathematics, basic sciences, engineering sciences and industrial aspects to meet specific needs. Many industrial processes need interactive and creative means to address the specific issues. Sometimes the engineering products are governed by standards and rules of varying proportion, depending on the disciplines. The economic, environmental and esthetic issues also surface in many cases. The graduates should acquire sufficient design skills through well balance Design related courses in the final four semesters. In addition students will be encouraged to take the elective courses from their own Departments as well as from other Departments. The elective courses have to be framed so that they enable the graduates to address new/ unsolved problems with imagination, courage and intelligence.

6.2.5 The Elective Courses, Open Electives and the Concept of Minor

The elective subjects develop the special talents of the individual students to serve the varied needs of society. These also provide flexibility of opportunity for gifted students. Some open electives are encouraged to prepare the students to cross traditional borders in their future careers. Graduates with knowledge of various areas of specializations will be well positioned to take advantage of interdisciplinary developments. The concept of 'Minor' may be introduced. As per the definition of 'Minor', the students may be provided with the flexibility to credit a sequence of three courses in the "open elective slots". These three courses could be from a specific department forming their Minor. However, no compulsion is desired to be imposed on the students to opt for a Minor. The scheme should remain voluntary.

The curriculum for the remaining eight semesters is recommended to have the substructure as shown in Table 6.2.

Table 6.2

Type of Courses	Credits (/ Courses)
1. Compulsory Professional Courses	36 or 9 courses
2. Elective (Departmental) Courses	20 or 5 courses
3. Open Electives	12 or 3 courses
4. Departmental Laboratory Courses	08 or 2 coures
5. Humanities and Social Science	04 or 1 course
6. Baccalaureate Project	08 or 2 courses

The final four semesters will have about 22 courses, i.e. equivalent to 88 Credits. The total Credit requirement for the Baccalaureate degree is about 168. The above mentioned break-up is a sample. The colleges/ institutes will have enough freedom to add or subtract the compulsory courses. Similarly, the interchangeability between departmental elective courses and the open elective courses should be allowed.

6.2.6 Baccalaureate Project

One of the most important activities of the practicing engineer is the solution of problems that arise in converting physical principles into useful devices. This involves creativity, imagination and reasonably strong analytical ability. These qualities can be nurtured and developed during the formative years of engineering education. Realizing the importance of training engineering students to think creatively at the earliest stage of their holistic development, Project Work is required to be introduced. The students should be encouraged to select projects of manageable size and complexity. Preferably such projects are to be executed in a group-wise manner. The project should begin with the submission of proposal generated by the students for authorization of the faculty-mentor.

Table 6.3 provides the number of courses under different categories in a model curriculum.

Table 6.3: No. of suggested courses in different categories

	Phy	Chem	Core lab	Math	ESc (C)	ESc (O)	HSS	EP	OE	Dept	Total
No. of Courses	2	1	1	3	3	4	3	2	3	18	40
Weight	5	2.5	2.5	7.5	7.5	10	7.5	5	7.5	45	100

ESc (C) = Eng Science Compulsory, ESc (O) = Eng Science Optional, OE = Open Elective, HSS = Communication Skills, Humanities and Social Sciences, EP = Engineering Practice.

6.2.7 Some Futuristic Thoughts

The paradigm of Engineering Research is on the verge of third Industrial Revolution. It is becoming increasingly clear that in the future the machines and devices will be guided substantially by the principals of life sciences. Quite a few new subjects have started emerging. **The Bio-inspired interdisciplinary subjects, such as, Biomimetics, Microfluidics, Microsystems technology, Bottom-up fabrication, Bioenergetics etc. are emerging fast.** These subjects require being included as the Department Electives or the Open Electives in all disciplines. Awareness programmes should be initiated so that Institutions and Universities start playing a proactive role in the inclusion of such subjects in the curricula.

6.3 EVALUATION MECHANISM

6.3.1 Continuous Evaluation

The students will be encouraged to be regular if a truly continuous evaluation system is evolved. There has to be a mid-semester examination which will carry at least 25 percent weight in the evaluation. The teachers are encouraged to conduct three to four quizzes at a regular interval. The quizzes together with the assignments are expected to form another 25 percent weight of the final evaluation, leaving about 50 percent weight for the end semester examination. The examination philosophy needs to be overhauled to encourage conceptual understanding of the subjects and creating a fear-free environment.

6.3.2 Grades and Grade Point

At the end of the semester, a student will be awarded a letter grade in each of his/ her course by the concerned Instructor-in-Charge taking into account his/ her performance in the various examinations, quizzes, assignments, laboratory work (if any), etc., besides regularity of attendance in classes.

In this section, a sample is being presented in order to explain the methodology of awarding the letter grades. The colleges/institutes may apply their own philosophy for the evaluation mechanism. The sample should not be viewed as being prescriptive. In the sample, nine letter grades have been used. The letter grades and their numerical equivalents on a 9-point scale (called grade points) are shown in Table 6.4.

Table 6.4

Grades	Grade Points	Interpretation
A+	10	Outstanding
A	9	Excellent
B+	8	Very Good
B	7	Good
C+	6	Average
C	5	Below Average
D	4	Pass
F	0	Poor
I	—	Incomplete

Note: F is failing grade.

6.3.3 Grade Point Average (GPA)

The Grade Point Average (GPA) is a weighted average of the grade points earned by a student in all the courses credited and describes his/ her academic performance in a semester. If the grade points associated with the letter grades awarded to a student are $g_1, g_2, g_3, g_4,$ and g_5 in five courses and the corresponding credits are $c_1, c_2, c_3, c_4,$ and c_5, the GPA is given by

$$GPA = \frac{c_1 g_1 + c_2 g_2 + c_3 g_3 + c_4 g_4 + c_5 g_5}{c_1 + c_2 + c_3 + c_4 + c_5}$$

6.3.4 Cumulative Grade Point Average (CGPA)

The Cumulative Grade Point Average (CGPA) is indicative of the overall academic performance of a student in all the courses registered up to and including the latest completed semester. It is computed in the same manner as the GPA, considering all the courses (say, n), and is given by

$$CGPA = \sum_{i=1}^{n} c_i g_i \Big/ \sum_{i=1}^{n} c_i$$

If a student is permitted to repeat any course, the new letter grade will replace the old letter grade in the computation of the CGPA.

6.4 CREATIVITY/THINKING OUT OF THE BOX

It is recommended that the curriculum accommodates the following during first two years of the Bachelor's Programme.

- An individual design project

- A group design project
- A monthly seminar
- Selected readings
- Measurements and data handling exercises
- Development of communication skills.

The design projects are expected to translate physical principles into useful device. This involves creativity and imagination as well as analytical ability. The monthly seminars are presented to the entire group of students in a year by experts from industry and the university. Subjects covered may include the following:

- An introduction to Engineering;
- Engineering Problem Solving;
- Creativity;
- Problem Solving during execution of a project such as Metro Rail;
- Written communications;
- Entrepreneurship;
- Project Engineering in Industry;
- Oral communications;
- Marketing;
- Economics;
- Patents and their implications;
- Case Study—for example "Indian Moon Mission";
- Case Study—for example "Technology of Mobile phones".

The topics are chosen to provide useful background in support of engineering experience from the practicing engineers.

Selected reading is suggested from the texts designed to introduce the beginners to a number of important topics, such as history of engineering, socio-technical problems, team effort, career in engineering, responsibility of the engineer, engineering design, the reduction and presentation of data and communication skills.

The measurement and data handling exercises may be designed to introduce the student to physical measurements and the proper presentation of data from the point of view of significant figure, graphical representation, and elementary statistics.

There should be some assignments that are "open-ended". The fact that always a better solution may be obtained having applied one's mind more carefully on the problem, the exercise will enhance the dormant creativity of the individuals. Given a proper ambiance, most students are expected to rise to the challenge and resolve to do the best job they can with the available resources.

Another useful suggestion is introduction of one **"Self study"** course during the senior years. The good students will be encouraged to credit such a course in lieu of an elective course. The student will be given freedom to choose any subject of his choice and then under the mentorship of a faculty member he/ she has to complete the study that was approved by the mentor at the time of registration. The focus of the self study should be on increasing flexibility within the mainstream system yet allowing the students to utilize their the dormant capacity through innovative ideas.

The weightage of Basic Sciences and various other components in the model framework of the curriculum are influenced by a historical report of the American Society of Engineering Education [Grinter (1)].

REFERENCES

1. L.E. Grinter, "Report on evaluation of engineering education (1952-1955)." *J. Engineering Education*, Vol. 46, No. 1, 1955, pp. 25-63. Reprinted in *J. Engineering Education*, Vol. 83, No. 1, pp. 74-95, 1994.

2. E.P. Torrance, Creativity in the Classroom, National Education Association. Washington, D.C., 1977.

3. Engineering Undergraduate Handbook, Fall 2007, Cornel University College of Engineering, USA.

7

Postgraduate Education and Curriculum

"Whatever you are doing, put your whole mind on it. If you are shooting, your mind should be only on the target."

— Swami Vivekananda

7.1 OBJECTIVES OF POSTGRADUATE EDUCATION

One of the major goals of the postgraduate study is acquiring special knowledge in a field that leads to increased expertise in a narrow area. However, many educators feel that a broader conception is an extremely important component of postgraduate education. The student who studies intensively in one particular field possesses a ready skill in the special area which may have immediate utility. Such expertise has only a temporary value without the overall intellectual growth through which the individual can master new techniques in any of the numerous related fields. The acquisition of techniques is, therefore, incidental in postgraduate educational experience. The deepening of insight and understanding, and the development of the stronger intellectual and scientific foundations are required for reaching excellence in the field involved.

Hence, the objectives of postgraduate study in engineering are the development of (i) a more general and fundamental understanding involving mathematical and natural sciences; (ii) more general and more powerful methods of analysis; (iii) capacity to understand the advanced work, through which the field is advancing; and (iv) imagination, and intellectual capacity to make new advancements.

7.2 REQUIREMENTS FOR A STRONG POSTGRADUATE PROGRAMME

In the year 2001 the Ministry of Human Resource and Development (MHRD) circulated its "policy framework". The policy framework was formed on the basis of the recommendations of Rama Rao Committee. The Committee, headed by Prof. P. Rama Rao was constituted in September 1995. The committee submitted its Report in 1999. Finally in 2001 the MHRD took it up as the "policy framework". The salient features of the "policy framework" were the following. The duration of MTech was recommended to be increased from 18 to 24 months to strengthen project work (research component). Graduate Aptitude Test in Engineering (GATE) examination was recommended to be continued and it was thought that National Co-ordination Board of GATE would be constituted once in three years. Self-financing students were to be encouraged for admission in branches with greater demand. The scholarships for MTech were to be raised from Rs 2500 to Rs 5000 per month. Greater focus was emphasized on course content and choice of new and emerging areas with the help of an AICTE standing committee reviewing these every three years. Accreditation was suggested as a tool for quality management. Concept of 'Technology Entrepreneur' was encouraged.

The essential requirements for a strong postgraduate program are the following: (i) Specially qualified faculty, (ii) Students of superior ability, and (iii) Adequate administrative and financial support. The 'policy framework' was in support of the requirements. Without each of these requirements, culmination of a successful postgraduate program is difficult.

7.3 THE POSTGRADUATE FACULTY

Dedicated and outstanding faculty is the most important requirement for any successful postgraduate program. Good faculty must have drive, and sincere interest in the development of manpower. The postgraduate faculty selection may be guided by the following. The selected faculty member should possess (a) creative talent and receptiveness to new ideas, (b) fundamental and critical understanding of one or more fields of engineering (c) ability to relate knowledge and experience in one field to a total concept involving many related fields (d) ability to support the graduate program through external funds (e.g., from industry) that also serves the dual purpose of integrating industrial needs.

Good postgraduate faculty must have desire to transmit his ever growing knowledge and understanding to young people. The strong faculty is composed of a group of academics of diverse talents and interests who are dedicated to the overall objective of providing the stimulation and environment for the professional growth of themselves and their students.

The programmes that would be offering postgraduate degrees must have adequate faculty members and each one of them should be well qualified to teach postgraduate courses. The faculty members have to be active in research and development (R&D) activities. They should have some sponsored/ industrial research projects. Securing patents is highly encouraged. The R&D activities will be substantiated by writing papers and (or) books. The faculty members should be capable of developing and floating new post graduate courses in the emerging areas. The country will need huge manpower during next ten years to sustain the academic growth. The focus will be to train at least some of the postgraduate students as the teachers. The postgraduate students should be encouraged to assist the teachers in undergraduate tutorials and laboratories. Having completed the programmes, the postgraduate students should be able to demonstrate teaching abilities.

7.4 ETHOS, PHILOSOPHY AND ANALYSIS

The following points are to be remembered and integrated with the philosophy of the College/ University/ Institute. A significant part of the published work from any Institute/ University comes as a result of the efforts to train postgraduate students. Faculty members should treat students as colleagues, whether in the classroom, office, or laboratory.

However, a postgraduate program does create its own set of challenges. There are two components of any postgraduate engineering education: Courses which are based on classroom teaching (course driven like MSc/MTech degree) and those which are research driven (namely MTech and PhD thesis).

Master's courses are extremely expensive in the use of staff time and, given the increase in undergraduate teaching loads, it is difficult for many departments to increase their intake of Master's students without an appropriate increase in staff. Usually the students have too few options for variety of electives. There is also a lack of concern among students in improvement of scholastic aptitudes. Lastly, postgraduate programs should focus on professional motivation and technology orientation based on socio-economic factors. This can be achieved by making these programs more interdisciplinary.

Thus the following **recommendations** can be made:

There should be emphasis on writing and communication skills. Having found that the intake to postgraduate programs is mostly from the private colleges in India, it is necessary to inculcate this along with a sense of international business environment. These would enable the students to be entrepreneurs.

Some flaws in the postgraduate education can only be removed at the UG level. There is an urgent need to teach basics of all sciences (Mathematics/ Physics/ Chemistry/ Biology) at the UG level. This would enable candidates to switch at least at the PG level. Many times, in the Indian system, candidates are unable to switch even if they have an interest. This also causes them to look for institutions abroad at the postgraduate level.

There is an urgent need to introduce professional motivational courses among the postgraduates. Several postgraduate students have little or no idea of what they might want to do after their education.

There is also a weakness in laboratory courses. The engineering component has to be strengthened in most laboratory courses at the postgraduate level. The same applies to experimental chemistry and physics. Students doing research in theory take as much as six months sometimes to learn computing basics.

More electives courses are to be introduced. The electives should cover almost all new emerging area of sciences and technology. This coupled with challenging problems for research will attract motivated students. The combined effect will offset the scarcity of master's students.

Better-managed interdisciplinary courses are to be offered. Future of research lies in emerging areas of high technology flowing from sciences like nanotechnology, biology, information technology etc. These postgraduate programs must act as forums. The activities should revolve around virtual departments and foster an environment for academic excellence. We would urge creation of as many as 100 such programs of which only 10% might succeed. IIT Kanpur took the lead in establishing the first interdisciplinary programme in Computer Science and Engineering in 1971. It flowered into a full-fledged department in all IITs in 1984.

There is a need to make patent oriented/product oriented peer influenced publication. It has been observed that several M.Tech theses do not get published in time. This fact should be discouraged. Each Master's thesis must be the source of at least one publication in a good journal.

One of the other ways to improve the quality of intake in postgraduate programs is to allow a mid-career research program from industry. There are a lot of government rules and regulations that make it imperative on the candidate from government research organizations to obtain prior sanction of concerned authorities before registering for a program at the university. This is a serious impediment. On several occasions the candidate is unable to secure such clearance within the time limit given to them by the university.

There is also a need to improve on the core strengths of the faculty so that patent/product oriented research may be carried out at the postgraduate level. There is a need to create an incubation center so that the product facilitation to industry is easier.

Our best qualified undergraduate students should repeatedly undergo a presentation of the opportunities available to them for immediate employment. They should have an equal opportunity to know and to examine carefully the available alternatives. Today's world the academic excellence can only be achieved though healthy post-graduate programs. All the departments should work together to adopt appropriate share of the following frontier areas:

- Energy Generation and Distribution;
- Communication Technology;
- Computational Mechanics;
- VLSI Design;
- Electronic Packaging;
- Biological Science and Bioengineering;
- Biomimetics;
- Water and Water Resources;
- Environmental Engineering;
- Advanced Materials Science;
- Petroleum Processing;
- Micro Electro Mechanical Systems;
- Thin Film Technology;
- Micro-scale Manufacturing;
- Modern Transportation Engineering;
- Nano-technology.

7.5 FUTURISTIC VIEWS

Today the principles of Biological Sciences are triggering plenty of innovations in cutting edge technology. Biomimetics has emerged as the most exciting scientific subject involving all disciplines. Biomimetics is aimed at design and fabrication of synthetic materials using the strategies adopted by living organisms to produce the biological ones. The organic/ inorganic composites synthesized by living organisms often display unique and desirable morphological, structural and mechanical properties, and represent informative models for the synthesis and design of complex functional materials. Biogenic materials are generally assembled from readily available materials, usually in aqueous media and at ambient conditions. Biomimetics approaches often involve the simulation of one or more of the different strategies used by organisms to selectively control mineral deposition for precise biological requirements. There is an urgent need to integrate different sub-areas of Biomimetics in various disciplines of postgraduate education.

7.6 CURRICULUM OVERVIEW

What are the components of the curriculum? How are courses sequenced? How do they fit together as a whole? A postgraduate program should be able to provide a summary of the objectives showing the course and credit requirements of the program, identifying required courses, electives, etc. How are practical experiences (labs) or internships or practical integrated with a postgraduate program? Is a comprehensive examination during the thesis defense required?

These are some open questions to be answered/considered while framing the curriculum for any postgraduate program.

The Master's degree is a 24-month postgraduate program with a beginning of research or design experience represented by Master's thesis. This degree prepares the graduate for an advanced position in the professional field, and provides a solid foundation for future doctoral study. The general requirements for the Master's degree can be at least 32 credits (8 courses) in coursework and at least 32 credits in research work. In most of the engineering departments, knowledge in advanced mathematics is an essential and mandatory component of postgraduate training. In addition, every department has some courses that orient an uninitiated student typically with the methodologies, practices and issues handled by the specific department. One such course is required to be credited by a student at the master's level. Every engineering department, such as Aerospace, Civil, Computer Science, Chemical, Electrical, Metallurgy, Mechanical etc. have number of specialized fields within the discipline itself. Some courses are to be credited that are specific and compulsory for such specialized fields. A student must be allowed to credit some postgraduate subjects relevant to the candidate's area of technical interest and thesis. These courses are elective courses and there should be sufficient number of elective courses available to the students. The students should be allowed to credit some elective courses from outside the department depending on their interest. Postgraduate research is expected to be guided by the philosophy of academic freedom and flexibility. Therefore, beyond a credit requirement from the compulsory slots (about 16 credits) no restrictions of courses are to be imposed. The students should have the freedom to shape their career and grow expertise that cuts across interrelated fields. The postgraduate programs especially should be well-suited to those who are enthusiastic to blend concepts and thoughts. Such academic freedom also promotes the sense of responsibility and allows an individual to think independently, act responsibly, and pursue research with self generated motivation and passion for learning.

REFERENCE

1. AICTE Report: "Reshaping Postgraduate Education and Research in Engineering and Technology", 1999, All India Council for Technical Education, New Delhi.

8

Improving Teaching – Learning Processes

"I believe that quality level is determined primarily by the actual design of the product itself, not by quality control in the production process."

— Hideo Sugiura

8.1 CURRICULUM

A comprehensive coverage on curricula has already been presented in the earlier chapters. In this chapter, we shall briefly mention about the issues related to curriculum. The Autonomous Institutions are required to specify and understand objectives of the group of courses that will be offered by any programme. The objective of the group of courses and their outcomes are to be analyzed in a systematic manner. The Affiliated Colleges are required to revisit the documents related to the basic university curriculum and the additions by the college. There is a need to look at the usefulness and relevance of the curriculum once in three years. Preferably this should be done by some competent group from outside the system. For autonomous institutes as well as for the affiliated colleges, it is desirable that the faculties develop new courses, devote adequate time and attention for teaching and evaluating the performance of the students.

8.2 EXAMINATION AND EVALUATION SYSTEM

The continuous evaluation system is a need for the systematic growth of mind. The continuous evaluation also brings about regularity of the students. For the autonomous institutions, mid-semester examinations, quizzes and assignments must carry significant weight so that the final examination does not play a massive role in deciding the grades. Laboratory courses are to be evaluated during the laboratory classes. For the affiliated colleges, internal assessments in the college should carry a significant part of the evaluation. The university examination should not carry the full weight for the purpose of evaluation. Usually in the university system, separate examinations are conducted for the laboratory courses. However, some weight is expected to be there for the regular laboratory classes as well. Notwithstanding the fact whether it is an autonomous institute or an affiliated college, the examination procedure should be overhauled to encourage conceptual understanding of the subjects and creativity of mind.

Educational psychologists developed a classification of levels of intellectual behavior important in learning. This became a taxonomy including three overlapping domains: the cognitive, affective and psychomotor. Six progressive stages of cognitive thinking, known as **Bloom's taxonomy**, are:

(i) Recall, (ii) Comprehension, (iii) Application, (iv) Analysis (v) Synthesis and (vi) Evaluation.

Cognitive learning is demonstrated by knowledge recall and the intellectual skills: comprehending information, organizing ideas, analyzing and synthesizing data, applying knowledge, choosing among alternatives in problem-solving and evaluating ideas or actions. This domain on the acquisition and use of knowledge has to be predominant in the majority of courses. The testing procedure should be well designed to test all the aspects of cognitive learning.

The question papers of final examinations and internal tests along with sample answer books, laboratory manuals/instructions and reports should be analyzed periodically (may be once in three years) and a report should be made available to the faculty members.

8.3 BASIC INFRASTRUCTURE

There should be adequate number of the class rooms; with proper air circulation and lighting. The teaching aids, such as, blackboard, multimedia projectors, computers etc. are to be modern. The faculty rooms should have availability of computers and internet access.

Size of the laboratories should be large enough to accommodate the students per section. The number of students per experimental facility should not exceed four. All the equipment should be in working condition to run experiments smoothly. Laboratory manuals; list of experiments etc must be well arranged.

8.4 ACADEMIC ADMINISTRATION

The administrate structure should be hierarchy free and horizontal. In managing the increasingly complex affairs of such Institutes, the Director (Vice Chancellor) should be supported by one Deputy Director (Pro Vice Chancellor) and the Deans. The support services (Works Departments, Finance, Health Care etc.) should be strengthened with professional expertise. All the consultative and advisory bodies of the Institute, such as the faculty council, student council, staff council etc. should be well defined and structured in a way to promote the Institutional goals and objectives free from any bias and overtones.

The Academic Calendar should be drawn in advance and it should be announced at the beginning of the year. Any deviation from the Academic calendar must have approval of the Academic Senate. The final result should be published within one month of the completion of the end-semester examination. Under no situation, there will be any interruption of scheduled lectures, laboratories and examinations.

Practical measures have to be adopted to streamline the organizational structure and improve the overall quality of its faculty. Reforms in governance are essential in order to evolve a system that will encourage openness and introduce transparency at every stage of academic procedures.

8.5 STUDENT CARE

The courses should be taught to provide integrated learning. The philosophy has to be to arouse the inquisitiveness of mind and encourage the students to probe deep into the concepts. Emphasis should be on learning by reasoning. Some very important suggestions for the educators are the following.

(a) The faculty members submit the course files at the end of each semester. The course files should include the lesson plans, the quizzes, question papers, special hand outs etc.

(b) Care must be taken to ascertain that the Laboratory courses are well designed and they clearly enhance the concept.

(c) It is highly desirable that students can present themselves well. Each student should be required to deliver at least one seminar talk in each semester.

(d) The students should be nurtured in such a manner that they understand the meaning of self learning in a true sense.

8.6 THE EDUCATORS

It is expected that the faculty student ratio is at least 1:15 and all the faculty members should have a post graduate degree. Even if any specific program becomes bound to recruit some faculty members with Bachelor's degree due to the pressing need of the program, those faculty members should be given opportunities to complete the postgraduate degree on a priority basis. There should be additional encouragement from the college authorities to have at least some faculty members with a PhD degree. The faculty members should have a decent track record. There has to be emphasis on having adequate number of faculty members in the Basic Sciences and Humanities. In many engineering colleges, the faculty members in Basic Sciences are overburdened with teaching responsibilities. It should be understood by the authorities that a well balanced faculty is key to the growth of the Institution. There should be formal and informal collaborations between the faculty members of the Engineering stream and their counterparts at the Basic Science and Humanities Departments.

The faculty members should not remain isolated from the developments that are taking place elsewhere. The following aspects will be overseen the administrative body

(a)   The faculty members visit other institutes in India and abroad, participating in national and international conferences.

(b)   The academic experts from other institutes must visit as the examiners, invited speakers, and the motivators

(c)   The experts from Industry or R & D laboratories should visit the Institutes/ Universities as the invited speakers and adjunct faculty.

(d)   Collaborations of the individual faculty members with other academic institutes/ industries / R & D laboratories.

8.7 LIBRARY FACILITIES

The libraries must have adequate number of titles in all the subject areas. The volumes per title have to be sufficient to meet the needs of undergraduate students, post graduate students and the faculty. The libraries should subscribe to the leading journals in the subject area nurtured in the college/university. The librarian should be well qualified and should have adequate background in Library-Science. It is desirable that the printing and reprographic facilities are created within library premises. Today, Membership of INDEST and networking with other libraries are essential components of a good library.

8.8 OBJECTIVES AND OUTCOMES

The objective of any programmme should be well defined. These should be derived from the vision of the college/university, its working environment, availability of resources etc. A successful programme may be aimed at fulfilling some or all the following objectives:

(a)   Abilities to apply knowledge of basic sciences, and engineering sciences appropriate to their disciplines.

(b)   Abilities to design and conduct experiments analyze and interpret data.

(c)   Design a system, component, or process to meet desired needs within realistic constraints such as economic, environmental, social, ethical, health and safety, manufacturability, and sustainability.

(d)   A proper understanding of professional and ethical responsibilities

(e)   Capability to function on multi-disciplinary teams.

(f)   Capability to identify, formulate, and solve engineering problems.

(g)   Abilities to appreciate the need for, and engage in, life-long learning.

(h)   Abilities to use the techniques, skills, and modern engineering tools necessary for engineering practice.

Similarly, a successful Engineering Programme is expected to result in some/ all the following student learning outcomes:

(a)   Graduates will demonstrate basic knowledge in mathematics, science and engineering.

(b)   Graduates will demonstrate the ability to design and conduct experiments, interpret and analyze data, and report results.

(c)   Graduates will demonstrate the ability to design a system or a process that meets desired specifications and requirements.

(d)   Graduates will demonstrate the ability to function on engineering and science laboratory teams, as well as on multidisciplinary design teams.

(e)   Graduates will demonstrate the ability to identify, formulate and solve the engineering problems pertaining to the discipline.

(f)   Graduates will demonstrate an understanding of their professional and ethical responsibilities.

(g)   Graduates will be able to communicate effectively in both verbal and written forms.

(h)   Graduates will have the confidence to apply engineering solutions in global and societal contexts.

(i)   Graduates should be capable of self-education and clearly understand the value of life-long learning.

(j)   Graduates will be broadly educated and will have an understanding of the impact of engineering on society and demonstrate awareness of contemporary issues.

(k)   Graduates will be familiar with modern engineering software tools and equipment to analyze engineering problems pertaining to the discipline.

As such, the accreditation procedure is envisaged to analyze the "Objectives and Outcomes" of every programme. The above mentioned sample 'Objectives and Outcomes" are in line with the **"Washington Accord"**. A comprehensive overview on this issue has already been covered in Chapter 2. In brief, the issue can be summarized as **"Say what you do"** and then **"Do what you say"**.

8.9 THE EDUCATIONAL MISSION

The Institute has to spell out specific missions depending on the available intellectual and physical resources. The missions will be formulated based on societal need as well. The missions will focus the academic program to cater the demand of the societal-economy and industry. They will be continuously revised and upgraded to meet the expectations of the beneficiaries. Quality of education lies in the depth and breath of the student's knowledge and skill to apply that knowledge in problem solving. The challenge to the Institute is to identify the optimum learning ambiance, creative spirit and novel teaching methods.

Every Institute/ University should be able to make a critical assessment of own strengths and weaknesses over a period of three years. There should be sincere efforts from the Institute/ University to correct the weaknesses.

8.10 GENERAL RECOMMENDATIONS

The end-semester examination should not carry more than 50% weight. The mid-semester examination, quizzes and assignments should carry 50% weight. Emphasis should be there to follow continuous evaluation system. The degree should be awarded on a credit based approach.

The first four semesters should cover mathematical, physical, chemical and life sciences. There should be coverage on environmental sciences. The "humanities courses" must carry at least 7-8 % of the total credit requirement.

Besides compulsory "engineering science" courses, there should be some optional courses in "engineering sciences" in order to provide overall exposure to various branches of engineering.

During the fifth semester, a student will be encouraged to select one area in the other streams (Minor) and if he/ she earns credits for three courses (including departmental and open electives) in that area, the student will graduate in his "Major" together with a chosen "Minor".

The students should undergo courses designed to improve "skills" such as software skills, accounting, hardware skills and instrumentation.

There should be an effort to blend the subjects, such as, biotechnology, bioinformatics, genetics etc. with the mainstream subjects of various streams.

There should be significant weight for the laboratory courses and field work. The laboratory courses should have some challenging experiments where a student would be required to design and fabricate the experiment. The laboratory facilities should be motivating for learning and taking up engineering as profession.

Emphasis has to be there to develop all-round communication skills. Students should be taught to nurture a wide range of communication skills, staring from presenting experimental data to presenting any theme or concept verbally.

There has to be events such as seminars and group discussions as the components of the credit.

REFERENCE

1. ABET 2000-CRITERIA FOR ACCREDITING ENGINEERING PROGRAMS Effective for Evaluations During the 2002-2003 Accreditation Cycle onwards, Accreditation Board for Engineering and Technology, Inc available at: Website: http://www.abet.org.

9

Governance and Management of Technical Institutions

"The governance of higher education in the 21st century needs to develop a fusion of academic and executive capacity rather than substitute one for the other."

— OECD Report

9.1 BACKGROUND

India has the world's third largest stock of technically and professionally trained human resource. Professionals and technologists educated in India are respected and are in demand all over the world. The remarkable growth of the Indian economy during the last several years is to a great extent due to the engineering and knowledge-based industries and services. India now aspires to become an economic power house and a Knowledge Capital of the world. Can this dream be translated into reality in an India which is characterized by extremes and contradictions in almost all spheres of human activity, including higher education in general and technical education in particular?

India has one of the largest technical education system in the world with undergraduate intake capacity already exceeding 8 lakhs per year. But, only 15-20 % of our engineering graduates are employable, as per NASSCOM. And despite this massive base of the pyramid, we produce no more than 1000 engineering and applied science PhDs per year. Compared to 12% contributions by China to scientific research publications in the world, India contributes less than 3%. The percentage of high impact research papers from India is less than 1% of the global science. As per Mr. Narayana Murthy of Infosys, the number of patents per one million people is one in India and 289 in the US. The number of researchers per one lakh population is 149 in India and 3805 in the US. Further, nearly 150 new companies come up each year in USA due to the efforts of the faculty and staff of MIT alone. Nearly 4000 such companies in the US have a combined sale worth $232 billion. This means that the productivity of every MIT graduate is equivalent to the productivity of nearly one lakh Indians.

Schools, universities, and other learning institutions worldover now encounter far more challenges, and are subjected to an unprecedented level of external scrutiny. The growing concern for value for money, efficiency, excellence, greater access, quality assurance, competitiveness, accountability, etc has altered the way the education is being managed in developed countries. Universities in developed countries nowadays are much more governed by market ideologies and the corporate culture of efficiency and effectiveness. The change in governance ideology in the higher education sector has undoubtedly altered the ways higher educational institutions are managed. Changes to control ways of monitoring, assuring and assessing the quality of education are universal requirements. Increased accountability has driven education practitioners and academics to engage in devising, and searching for, different mechanisms and strategies in order to assure quality, particularly with more weight being given to satisfy the three major stakeholders in the system.

That our higher education is not doing well is publicly acknowledged by our highest policy makers. The fact is that the main responsibility for the sad situation rests on the ministry of human resources and development (MHRD) itself, along with the state education ministries, because of their politically motivated, caste centered, illogical and corruption ridden policies and rules and regulations for governance and management of our academic institutions in the whole country. Good governance means framing flexible and pragmatic rules and regulations under the guidance and supervision of inspiring leaders in the governing body which will enable efficient manager(s) and chief executives of academic institutions to translate the mission and vision of the institution into a reality. The role of a government, state or central, and their agencies should be to hold hands of academic institutions and to facilitate dissemination and creation of highest class knowledge. In practice, however, our management system for educational institutions continues to be archaic.

9.1.1 Chaotic System

Unity in Diversity continues to be our national obsession and thus we have one of the most diverse, albeit chaotic, technical education system in the world. At present (Year 2008), the technical education system includes 12 IITs as Institutes of National Importance, one IISC, 22 National Institutes of Technology (NITs), 21 IIITS, 24 Central Universities with engineering faculties, over 200 affiliating/ non-affiliating / self-financing State Universities which include over 20 Technical Universities, and numerous Agricultural, Pharmacy, Architecture and Medical Universities. Also included are over 150 private Deemed, De-Novo - Deemed (read "*Instant*") and state Universities, 18 Open / Distance Learning universities offering technology programmes, over 30 public and private "Directed/ Dedicated" autonomous Research Centres as well as several government departments (such as DRDO, ISRO, DAE, BARC) and associated R&D institutions recently empowered with Deemed University status, and over 2000 private technical colleges distributed asymmetrically all over the country. Further, the MHRD has just announced its plan to the creation of 8 new IITs, 20 new IIITS, 12 new Central Universities, 10 new NITs, 2 new Schools of Planning and Architecture just before the national elections.

Higher education in India is on "concurrent" list which means that the responsibility of providing the education is divided between the state and central governments with the inevitable consequence of little or no coordination among themselves. Central institutions are set up through an Act of Parliament, Deemed-to-be universities(both pubic and private) through a UGC Act, state universities, public, private and public-private ones, through an Act of the State Legislation, and Engineering Colleges through a plethora of approvals from different state and central bodies. A private university bill is awaiting consideration by the central government for over two decades. As yet, there are no clear views of the government on foreign universities which are interested in operating in India. Ironically, many approved private institutions in India are already running over 100 collaborative UG programmes on a twinning basis with foreign institutions, with as well as without the approval of AICTE.

The All India Council for Technical Education (AICTE), first set up as an Apex Advisory Body by the central government was later empowered in 1988 with statutory powers to plan, formulate and maintain norms and standards, to assess quality through accreditation, fund priority areas, monitor and evaluate, maintain parity of certification and awards, and ensure coordinated and integrated development and management of technical education in the country. Declared as "autonomous', AICTE has little autonomy in practice since its administrative and financial controls rest with the MHRD. The AICTE Act does not provide clearly defined legal empowerment to act against defaults and deficiencies of any academic institution. It is also now subservient to the decisions of another Apex Body, namely University Grants Commission (UGC) which, in principle, should have no jurisdiction in matters of creation of Deemed Technical Universities and other aspects related to their academic programmes and performance. Consequently, most of the AICTE activities are centered on cycles of approvals and quotas and shoddy accreditation, leading to large scale corruption in the whole system. The situation is worse when it comes to the approval cycles for Deemed University Status in which UGC and MHRD are the major decision makers. Since most of the private technical colleges and universities are owned and controlled directly or indirectly by politicians, reforms in the governance system are the last priority of any government.

Besides UGC and AICTE, India has created another 12 statutory bodies such as Council of Architects, Pharmacy Council of India, Indian Council for Agriculture Research, Indian Council of Medical Research, etc, which deal with or influence in one way or the other the processes of recognition, benchmarking, or registration of professionals in technical education. Some of these bodies have brought out status reports highlighting the weaknesses of related areas of technical education. But, it is almost impossible to bring these bodies on a common platform to interact with each other and present an integrated view to the government for radical reforms in the governance and management of our institutions.

A large number of excellent status reports on higher and technical education, the latest being that of the Knowledge Commission, have been submitted to the government at various times by much respected educationists of the country. All these reports have emphasized time and again the need to reform our educational system to provide flexibility, respectability, autonomy, accountability and least involvement or interference from any government, centre or state, body. A model governance framework must provide academic, administrative, and financial autonomy with built-in measures for checks and balances and accountability to usher in a culture of *"institute of academics, for academics, by academics"*.

9.2 ROLE OF PROFESSIONAL BODIES

Quoting the words of the National Knowledge Commission, "most engineering graduates do not possess the skills needed to compete in the economy, and industries have been facing a skills' deficit". The present engineering education and the skills imparted are not adequate enough to meet new and emerging challenges of internationally acceptable equality of capabilities which is required for international mobility in the WTO environment. Further, in the absence of an Engineers Act, the engineering profession has no legal status in India. Since India is not a member of Engineers Mobility Forum and only a provisional member of the Washington Accord at present, recognition of our Engineering degrees abroad is not automatic.

Just how to balance rigorous academic training of an engineer versus preparing him for industry-readiness and further to enable him/her for international mobility is a matter of concern to both academic institutions and professional bodies such as institutes, societies, councils, and national academies, FICCI, CII and NASSCOM. Re-engineering engineering profession to meet the evolving requirements of the profession should be an important objective of these bodies. It is, therefore, paramount that professional bodies play a strong interactive role in the academic planning, regulation and assessment of the technical institutions. At present, these bodies play very little role.

Most professional disciplines have their own apex body which could interact with academic institutions, to offer advice and appropriate training courses for upgrading knowledge as also degree qualifications. An Engineers Bill has been drafted by the Engineering Council of India (ECI) which is an apex body of some 25 professional associations, societies, councils and institutions such as Institution of Engineers, Institution of Electronics and Telecommunication Engineers, Institute of Chemical Engineers, Computer Society of India, Consulting Engineers Association of India, etc covering almost all disciplines. The bill, presently pending with the Government of India, proposes a statutory body to register and license engineers. The ECI aims at becoming a representative of the professional associations which will accord recognition of professional qualification and experience, bring about discipline and accountability, and monitor ethical conduct among professionals.

9.3 GOVERNANCE OF ACADEMIC INSTITUTIONS

Let us look at the governance and management systems of various classes of institutions to understand what constrains and restrains them from quality performance.

9.3.1 IITs

- All IITs have been created by an Act of Parliament essentially following the 1961 Act for IIT Kharagpur. Despite shortcomings, neither the Act nor the Statutes have undergone any change since then even though several committees of IITs have jointly proposed changes necessitated by the new education challenges. No government has considered it important enough to go through the hassles of getting it changed by the parliament. Some of the rules and regulations are so archaic and outdated that IITs had to change them per force individually so as to be able to run the institutions effectively. As an example, Deans play an important academic and administrative management role in each IIT and yet the position of a Dean does not exist in the statutes even today.

- Presently, the Board of Governors (BoG) is the apex body which is appointed by the government (read "minister"), except for the two faculty representatives. The Board has very little autonomy in financial and administrative matters since all such activities of the institute are regulated by the rules approved by the MHRD. With the exception of a few past Chairmen of IITs, most have considered it as a decorative position and have done little or nothing for any process of development of the institutes partly because they have little time, and partly because they have little understanding of what is required to nurture an academic institute to an international level. As a general experience, most members of the Board, do not contribute much to the deliberations of the board.

- Recently, MHRD has made it mandatory even for IITs to have a senior officer of the ministry on the BoG whereby reducing its independence.

- Purchase procedures are normally complicated enough and now have been made inflexible and threatening because of the overriding guidelines issued by the central vigilance office. As a result, most faculty members shy away from purchase processes and thus most of the budget has to be spent somehow in the last few months of the financial year.

- The only autonomy IIT faculty enjoys is "what to teach, how to teach and how to assess". With recent imposition of additional reservations and quotas enforced by the central government, even this autonomy for students and faculty has been compromised.

- The Senate is the Apex academic body of IITs and consists of all Professors which now number well over 250 in each IIT. Such a large body is invariably bogged down with routine matters and is unable to perform the role of an academic watchdog and a think-tank.

- Since the government decides all the administrative, financial, appointment, and promotional rules, the IIT administration can at best follow the rules robotically, without making any efforts to be innovative to attract the best.

- A Council, the Apex Body of all IITs, is headed by the Minister, HRD and consists of Directors and BoG Chairmen of IITs, and a variety of other professionals and political representatives. All decisions affecting all IITs such as admission process of UG students, tuition fee, scholarships/fellowships, reservation quotas, salary scales, promotional avenues, starting/closing down or renaming a BTech programme or a Department, creation of a corpus fund, etc are all within the purview of the Council which unfortunately meets only rarely.

- Fortunately, since the post graduate (PG) programmes were not spelled out in the IIT Statutes, the status and health of MTech, PhD and research programmes in IITs is of little interest to the Council. Nevertheless, the Council exercises its control through the process of approval of fellowships and financial commitments for running such programmes.

- Although the IITs have processes in place for self-assessment of the faculty, assessment at the time of promotion, and also assessment by the students of the teaching-learning process adopted by the faculty, there is little or no overall assessment of the outcomes of significant outlays for R&D in these well-endowed institutions, except for a general report of a Review Committee set up by the MHRD every ten years or so.

- The appointments of Chairmen and members of BoG, and the Directors of IITs are increasingly becoming politically influenced. Filling of a vacancy of a Director by the process of selection through an humiliating interview chaired by the minister, HRD may take as long as a year . Headless IITs or one IIT Director heading simultaneously one or more IITs or other national institutions are common features at present.

- IITs spend a lot of time on conducting financially lucrative JEE and GATE for UG and PG admission. Whether these examinations are the best way of testing the aptitude of the candidates and whether the IITs should spend so much time at the expense of their own R&D activities has been debated. However, vested interests do not allow meaningful discussion on the subject; leave alone consider the possibility of handing over these responsibilities to an autonomous professional organization.

9.3.2 NITs

All NITs heaved a sigh of relief after being rescued from being primitively managed by technical education departments of the concerned states. After a few years of being in doldrums due to lack of any strategic planning, these institutions have now been assured of enhanced funding and better working ecology by the central government and are now expected to be the next to the level of IITs. Although the Act and Statutes for the NITs have been approved, these have yet to be promulgated. In the transition stage already lasting several years, the governance and management style of the IITs is being implemented in a limited way.

Instead of overcoming the deficiencies of the IIT system, the governance is being more tightly controlled by having two MHRD and two State government officers in the BoG and other bodies of the institute. Many Chairmen of these NITs lack management skills in an academic or knowledge ecology. With the present governance set up providing little or no administrative and financial autonomy, management of many of these institutions is already in turmoil. With limited academic autonomy, a plethora of reservations and quotas for admission of students, interference by the state governments through various mechanisms, how these institutions will reach any level of UG and PG closer to that of any IIT is anybody's guess.

9.3.3 Central Universities

Central Universities, huge in terms of various faculties, with the exception of those in the Northeast, are well endowed with financial resources and infrastructure. The engineering departments of most of these universities have now been given the status of Institutes of Technology (IT). The style of governance and management and the autonomy enjoyed by the ITs is controlled by the Vice Chancellor and thus is no different or better than that of any other department of the university. In principle, the VC has more autonomy in functioning than the Director of an IIT has. However, due to very large membership of very heterogeneous group of members of such bodies as the Court, Academic Council, Executive Committee, etc, the VC has little freedom or inclination to be an effective leader of the IT unless he commands an exceptional academic and administrative respect. As a result, large universities, with some exceptions, are not very successful in nurturing high quality engineering post graduates and researchers. Although each central university in the Northeast has its own problems, the lack of infrastructure, turbulent political and social environment, and the difficulty of getting reasonably qualified professional faculty are daunting challenges common to all of them.

9.3.4 State Universities

State universities are totally at the mercy of the governor, minister of higher education and officials of the department of education of the state. Appointment of the VC, and faculty are controlled by the state. In some states (such as MP), the Registrar and Finance Officer are deputed by the state and report directly to the state government rather than the VC. Getting funds on time even for salaries or approvals for any new academic activity are herculien tasks. State universities have no choice but to learn to manage academic activities with very limited facilities, insufficient funds and continous political interference. Some state universities have been allowed to run self-financing courses which provide additional funds and some freedom to act. On the whole, quality of governance and education at any level leaves much to be desired.

Most State universities are also charged to look after the overall academic activities of a large number of affiliated colleges. Some States have also set up Technical Universities specifically for the purpose of affiliating engineering colleges. These universities are financially rich since they earn substantial money from the fees from the affiliated colleges by conducting examinations which in most cases is their main academic activity. However, some Technical Universities have started running UG and PG programmes. These universities, like other state universities, have very limited autonomy in functioning and very few faculties to nurture any respectable post graduate programme.

9.3.5 Deemed Universities

A university status for a privately funded institution through a private Trust or Society provides, in principle, total academic, administrative and financial autonomy to the governing body set up by the Trust. In practice, however, this means the "owners" of the university assume the controlling positions such as Chancellor, Vice Chancellor, Registrar, Finance Officer, etc. Though such management is generally very authoritarian or even dictatorial, and profit oriented, it can, in principle, provide quality education if it so chooses. In practice, however this is not so at present in over 80% of the private Deemed universities. Nevertheless, dictated by the competition of attracting more students, some

15-20% of these universities are now making efforts to provide quality education.

Every private college, new or old, aspires to obtain university status since it allows it to admit as many students as it likes through autonomous procedures, teach whatever it wishes, conduct assessments and examinations as per its own norms, charge additional fees in many ways, and hire, fire and pay faculty as per its sweet will. Thus, such a commercialized university has good avenues to mint a lot of money. Though there do exist UGC /AICTE guidelines for academic and administrative norms for good practices and quality governance but these are followed on paper only since there is little credible monitoring by any agency. For example, a University is required to have PG, PhD and research programmes in its major academic departments. Most of the private universities cannot sustain these programmes since they have poorly qualified and inadequate faculty, poor infrastructure and facilities for costly PG studies. Clearly, most private colleges have used the route of corruption to get required approvals for a university status, without having satisfied norms for basic academic facilities.

The MHRD has also provided another route to obtain DeNovo-Deemed university status by any institution which is willing to start teaching programmes in frontier areas such as Biotechnology and Information Technology. Many universities have been created this way, sometimes without any buildings, faculty or infrastructure.

The government of India has recently conferred university status for granting PG and PhD degrees to some 22 GoI funded research institutes / organizations of various government departments and several other private research centres. Set up to conduct high quality "dedicated" research as their main mission, these "one subject" universities are an anomaly in a knowledge society since these lack the ambience of an academic campus, liberal education of a university, and qualified teaching faculty. Thus, the existing stimulating avenues of providing post graduate education and training through established academic institutions to engineers and research scientists working in such research centers has been demolished by the government without any academic debate.

9.3.6 Engineering Colleges

Most engineering colleges are privately owned and administered and are affiliated with the State universities. Over 85% of the engineering graduates today are the product of these colleges. The affiliated colleges have no academic autonomy since all academic matters related to admission of students, curricula and examinations are decided and dictated by the affiliating state universities, often without their active involvement in decision making bodies set up by the universities. Although the exact situation varies from university to university, very few colleges in some states have been given the Academic Autonomy Status on the basis of their infrastructure and academic achievements. In fact, many states have no rule permitting an "autonomous" status. Though guidelines for academic programmes in the colleges have been circulated by the AICTE, the affiliating university departments have imposed their own curricula and academic structures.

The private colleges are tightly controlled by the private Trust or Investor and the administrative and financial autonomy rests with the "owners". Since maximizing the profit is an all important goal for the owner, all academic and administrative actions are tuned to this philosophy. However, tough competition between colleges to get the best and maximum number of students has recently resulted in some improvements in the quality of faculty and the teaching-learning process in some (about 20%) colleges. Very few private colleges have started PG programmes for lack of facilities, faculty and potential for making money.

9.3.7 Private Institutions

Besides self-financing private colleges and universities, some new academic systems such as pubic-private partnership, private-foreign institution twinning programme, and NRI University have begun to emerge. Governance and management in such institutions are expected to be largely in the hands of the private partners. A transparent and rigorous regulatory and assessment mechanism still needs to be evolved for ensuring good management practices in such institutions.

9.4 SOME RECOMMENDATIONS

1. Today, knowledge driven education is evolving rapidly and requires global recognition and certification for globalized economies. There is a need to have a non-government, and autonomous Higher Education Body (consisting of representatives of prominent stakeholders) to act as a national think-tank for planning, benchmarking, forecasting and preparing guide lines for all aspects of a world-class higher and technical education in the country.

2. Roles of UGC and AICTE or another hybridized new organization need to be re-defined and confined to development of curricula, pedagogy, faculty, R & D temper, entrepreneurship, etc. These institutions should be empowered in their respective domains to ensure compliance with the rules and regulations.

3. All types of approvals for the establishment of new institutions and any additions or changes in the old institutions should be automatic, subject to certification of minimum requirements by designated certifying bodies drawn from professional institutions, councils and academies. Such information should be available as an "Open Access" document on a suitable website or a repository.

4. Credible accreditation as per international standards (e.g.; Washington Accord) should be handled by several autonomous non-government agencies drawn from professional and industrial organizations rather than by the existing NBA and NAAC which suffer from various serious infirmities. Such bodies need to be financially autonomous and self-sustaining. They should be empowered to display the assessment details for public scrutiny.

5. (a) An Act of parliament or that of state legislature for creating new universities should be a simple generic and general document in the form of a model Act consistent with the aspirations of a knowledge society. Since an executive order of a state government or MHRD can create a university, the need for processing through an Act of Parliament appears to be redundant.

   (b) All functional aspects of rules and regulations, or ordinances should have the provision that these could be re-visited, and revised if necessary, periodically by the governing bodies of the institution without involving any government department. A model document for guidance of all institutions would be a useful template to follow.

6. India needs a large number of universities and technical institutions of the highest class. Large investments required for such ventures need participation of private, public-private or public-private-foreign partnership organizations. Level playing field with the same policies, procedures and opportunities must be provided to any such venture. The private university bill which has been awaiting clearance for over two decades and also a foreign university bill need to be enacted.

7. All institutions, public or private, should be encouraged to raise resources ethically from legitimate sources. The existing fee structure in public institutions is illogical. It is generally agreed world over that at least 20% of the budget should be met through fees. Poor and weaker section students should be given freeships by the governments/ institutions concerned. Those who can afford it financially should be required to pay reasonable charges.

8. There is an urgent need for setting up a GoI funded Education Development Bank which should provide loans to students at very nominal rates of interest. Such a bank should be empowered to recover loans from students after they have got jobs.

9. An academic platform for discussion of common problems by like- minded institutions such as IITs, NITs and other institutions coming upto a good standard is desirable. The constraints imposed by a Council of IITs or other institutions is meaningless today. All premier institutions should be academically autonomous to compete with each other and other institutions in terms of admissions of students, quality of teaching-learning processes, R&D, consultancy, etc.

10.   (a) Any institute, public or private, should be given an opportunity to aspire for a dignified recognition such as "Institute of National Importance" or any other suitable title by attaining certain benchmarks of achievements as certified by suitable independent professional agencies.

(b) The composition and the functions of the governing bodies of technical institutions, both public and private, need to be reviewed and revised . Membership (including Chairman) should be drawn from a databank of well known educationists, professionals, industrialists and alumni who have a demonstrated stake in the system.

(c) Monitoring and grading of the performance of an institute should be a mandatory responsibility of the governance system.

11.   Senate or an equivalent apex body for all academic matters should be small and compact for effective planning, monitoring and decision making. Routine academic matters should be handled by respective autonomous boards or committees of studies set up by the Senate.

12.   Quality of faculty determines to a large extent the quality of the institution and its product, namely students. Any type of implicit or explicit reservation by law or on regional, religious or caste basis would destroy a good academic system and should be guarded against. Institutions should, however, be required to pursue affirmative action policy to ensure national homogeneity and gender equality.

13.   Performance incentives for faculty must be introduced. Faculty which performs well should be allowed additional increments, some perks, or honorarium from their sponsored projects, consultancy, or industrial work during vacation periods. Faculty should be allowed liberally to work with the industry in whatever capacity on an appropriate time sharing, cost sharing or retainership basis.

14   All institutions should be encouraged to set up STEPs, EDCs, or TBIUs which allow faculty to not only mentor young entrepreneurs but also to join the companies so created as stakeholders. All institutions should be allowed to establish a corpus to be used for hand holding and/or venture capital purpose to support the nascent ventures.

15.   Collaboration between institutions, public, private or foreign, for any legitimate academic and R&D activity approved by relevant bodies should be allowed without seeking any approvals from any government agency.

16.   Vice Chancellors and Directors should be invited to hold the position on the basis of recommendations of a search committees consisting of members drawn from an approved list of distinguished academics and engineering professionals. The government should play only advisory but no active role in the selection process.

17. All academic institutions should be treated as "not-for-profit" Trust/ Society which means they can make tax –free profits which must be ploughed back into the institute in whatever form, including a corpus, if it is so deemed.

18. Public institutions of national importance should be given a fixed grant to take care of all recurring expenses. Beyond that, they should compete with others to raise resources in variety of ways. This experiment has already been tried out in IITs and IIMs successfully but has been given up due to the political philosophies of the government of the day. The government should, however, provide incentives in terms of matching suitably financial resources generated by the institution so that the Institute can establish a healthy corpus.

19. All institutions should be required to post total information regarding students, theses, faculty, infrastructure, management, R&D, publications, budget etc. The information should be made available as per an approved format/ template on an open-access website, along with the grades given by an approved accreditation body. The information should be adequate to enable stakeholders, in particular parents and students, to assign their own quality index for the institution.

20. The need for one or more common tests held nationally per year by a professionally competent agency for admissions at UG and PG levels has been felt for a long time and has been recommended by the New Education Policy-1987. With the help and supervision of premier institutes like IITs, an autonomous and self-sustaining National Testing Service needs to be nurtured to replace the existing systems. Only such an agency can cope with evolving changes to develop a transparent web-based system.

21. All technical institutions should be encouraged to follow a reasonably uniform system of continuous evaluation, credits and semesters. Mobility of students to earn credits from different institutions or to change institutions should in principle be allowed even though its feasibility will remain limited in India largely because of large numbers. Nevertheless, this practice will motivate weaker institutions to improve to compete.

22. Presence of foreign students on our academic campuses would challenge our teaching – learning process, add to the cultural ecology, and earn us a global brand name. The government must relax, if not abolish, its controls on admission policies for foreign students and visa restrictions so that they feel welcome in our institutions.

10

Promotion of Research and Industrial Interaction

"Higher education and research are probably the most productive instruments of human and natural resources in the long run. The optimal utilization of precious and scarce resources, however, depends on an efficient system of policy-making and policy-implementation."

— Joseph S. Joseph

10.1 BACKGROUND

As per the UNESCO definition, learning to learn, learning to do, learning to be (useful) and learning to live together constitute the fundamentals of any good education. The advent of the 21st century has ushered in paradigmatic changes and shifts in our concept and practice of education in general and engineering education in particular. The driving force behind these changes (and others related to technomic-globalisation) is the "Knowledge Power" which is the power to create and innovate usable and exploitable scientific information for providing new, cheaper, efficient and sustainable solutions for the needs of mankind. It is a multimodal and multidimensional process based on a combination of a various sciences, engineering skills, technologies, marketing, manufacturing and management techniques. Evolving as an explicit rather than a tacit force, increasing rapidly and yet also becoming obsolete just as rapidly; knowledge is now a tradeable and marketable commodity protected internationally by IPR laws. The economic growth of any society is now increasingly dependent on "creativity through human resources, innovation through research and development, and capital through intellectual property rights". Consequently, besides the conventional practice of assimilation and diffusion of knowledge, the creation of knowledge through research and development must be an important and integral component of any science and engineering education.

10.2 WHAT IS RESEARCH?

Research is to conduct scholarly inquiry, to develop, create, innovate or discover something hitherto unknown yielding an intellectual outcome to enhance our knowledge of nature in its various living and non-living manifestations of processes and products, which may benefit the society in some tangible or intangible ways. Whether research by engineering scholars should be rigorous or relevant, basic or applied, business driven or technology/knowledge driven has been debated extensively. However, the generally accepted concept of linear chain connecting basic science with applied science to engineering and technology is no more taken to be valid in the prevailing knowledge era. Thus, for engineering education, a broader definition of research which involves scholarly investigation or inquiry, resulting in a product or process, or which leads to intellectually stimulating knowledge ambience is now widely accepted.

10.3 WHY RESEARCH?

Prof J C Ghosh, a former director of IIT Kharagpur, told a group of teachers that teaching without research is like a stagnant pool of water, which, after sometime, is likely to stink. It is now universally accepted that teaching and research must go together. Whether good teachers are also good researchers and vice versa has been debated often. It is, however, generally concluded on the basis of experience that good research correlates positively with teaching efficiencies and, further, good researchers are relatively more committed and inspiring teachers.

Research in various forms is of considerable interest to a faculty member for a variety of reasons such as:

- To update, broaden and enhance knowledge capital and analytical skills of the faculty.
- To enhance self-confidence and professional respect and recognition among peers at home and abroad.
- To stimulate creative teaching-learning experience.
- To become an inspiring teacher.
- To broaden and enhance knowledge capital and analytical skills of research students.
- To nurture research students with the spirit of working together on research problems in a group effort.
- To nurture and mentor students to become the next generation faculty.
- To be available as a "think tank" to serve country, industry and society.
- To create entrepreneurs to support the creativity chain.
- To interact with industry through research or consultancy on real life problems for mutual professional and financial benefits.

- To participate in national and international conferences for exchange of ideas and growth of knowledge.
- To undertake consultancy work for intellectual and economic benefits.
- To earn intangible benefits in terms of recognition,fellowships of academies, awards and rewards.
- To bring laurels to ones institution and thus to create its brand image differentiation.
- To create intellectual property for the nation.

10.4 WHAT TYPE OF RESEARCH?

A relevant and mission mode research is, of course, of great economic value to the society which pays for it. Nevertheless, any type of scholarly inquiry and research, whether basic or applied, theoretical or experimental, which leads to creating intellectual capital is not only desirable but vitally important for engineering education.

Broadly speaking, typical types of research are:

- Study of natural phenomena and their applications in any area of science be it physical, mechanical, chemical, metallurgical, biological, medical, agricultural, cross-disciplinary, multi-disciplinary etc, which may lead to refereed publications in respected journals, to new knowledge, to improve existing engineering practices, processes or products, patent disclosures, or a useful databank.
- Design/develop/fabricate, on a laboratory/pilot plant scale, processes, products, devices, equipment etc for teaching, research, industry or society at large.
- Interact/consult with industry, civic authorities and service providers to improve/innovate/create processes/products.
- Write stimulating research monographs and textbooks for students.
- Create inspiring and stimulating teaching-learning materials, equipment and techniques.
- Interact with industry through research and/or consultancy on result / mission oriented problems for mutual benefit.

10.5 PRESENT SCENARIO

Despite several comprehensive reports on technical education in the country in the recent past, no reliable and credible primary or even secondary data on the input and output of research or research oriented education in engineering institutions in India exists at present. The most recent Knowledge Commission Report and the Observer Research Foundation funded report of Professors Rangan Banerjee and Vinayak Muley (BM report) are some of the best efforts to analyze data based on commonsense extrapolations. However, the data is already outdated within the last few years due to rapid (over 50%) increase of the number of engineering colleges (largely private) to over 2000 and private Deemed-to-be/state technical universities to 150 in the year 2008. The student intake capacity of these institutions has exceeded 8 lakhs. Besides engineering colleges, education at degree level is also offered by many of the over 250 state and central universities, as also by 12 IITs, IISc, 21 IIITs, 22 NITs, some specialized institutions and even some polytechnics. Out of over 4 lakh engineering graduates during year 2006, national institutes like IITs and NITs contributed only about 3%. It is noteworthy that a majority (over 85%) of the graduates pass out from private engineering institutions.

Further, as per the BM report, IIT grads, in general, do not opt for MTech or PhD at IIT as almost all IIT graduates get decent jobs before they actually get their degrees. However, a large number of these graduates will sooner or later go abroad (primarily to USA) for higher studies. A small fraction of less than 2% will join post graduate programmes in Indian institutions, in particular IITs and NITs. A still smaller (about 1%) fraction thereof will join PhD programmes.. Top institutes like IITs and IISc contributed less than 1% of the engineering grads, 20% of the MTechs and 40% of PhDs.

Despite all these large numbers of our academic institutions and graduates, the annual number of engineering PhDs is less than 1,000 which is less than 1% of graduating MTechs in the country. It is very likely that this number of engineering PhDs also includes doctorates from several applied science and science-engineering interdisciplinary areas. It is noteworthy that, in the US, more than a third of all PhD degrees in science and engineering and almost half of all doctorates in mathematics and computer science are awarded to non-US students. Among the students between 1998 to 2001, Indians account for 10% of science and engineering doctorates of which biomedical, aerospace and computers engineering constitute 16.3, 12.8 and 7.7% respectively. These numbers have increased substantially in the last few years.

The preceding data make it clear that a very large majority of our engineering institutions have failed to evolve from undergraduate teaching to post graduate teaching and research at MTech and Doctoral levels. The vast majority of private engineering institutions are primarily first degree level ones and have made little effort to introduce research oriented programmes for economic reasons as also for lack of qualified faculty. Clearly, the engineering education system in India has been unable to attract the best engineering students for post-graduate studies. Even the IITs are unable to attract their own engineering graduates and postgraduates for their doctoral theses. The BM study shows that only 1% of the graduating BTech class of an IIT opts for an MTech and only 2% of the graduating MTech class opts for a PhD.

The IITs have failed to meet the objectives of research oriented institutions and to take up the role of "Think Tanks" as was envisioned by the founders. The percentage of faculty engaged in serious research in any of the IITs is around 30 %. An analysis (such as done by the Institute of Scientific Information (ISI), USA) of the research publications per faculty in IITs shows that it is around 1 as compared with over 6 of the MIT faculty. Even worse, the citation index and impact factors of the IIT publications places the IIT faculty well below the achievements of globally known technical institutions in the western world, as also China and South Korea. This situation has been prevailing despite the "publish or perish" rule for promotions in the IIT system. The situation about research in most other public and private institutions is too pathetic to talk about (see the recent report of Prathap and Gupta). The challenge for the IIT system, as also for NITs and Technical Universities, is to enhance their overall output considerably so as to be able to impact the engineering education system and its research contributions.

Premier institutions are also expected to contribute to patents and industrial consultancy practice. An interest in these activities is a relatively new phenomenon within the last two decades or so. The "patent before publish" advisory has encouraged some faculty in premier institutions to file patent applications. The total no of patents filed per year by all of our national academic institutions is around 100 as yet. The number of such patents which have been utilized by any industry, although not known precisely, is too insignificant to talk about. As regards consultancy, the practice hardly exists in most institutions except for IITs and a few other premier institutions. The number of faculty participating in consultancy services even in IITs is rather small, probably less than 10 %.

10.6 FUNDING OF RESEARCH

Funding for all R&D activities in the country is presently mostly by the several S&T departments of the government of India and is less than 1% of the GDP which is one third of what Korea spends, and certainly much less than what most developed nations spend. The Indian industry is known to contribute about 20% of the total available R&D funds but these are primarily for their own in-house activities and are much smaller than what industry spends in developed countries.

A significant part of the government funds for R&D is being given to public engineering institutions through a whole variety of schemes of various S&T funding agencies/departments/ministries of the government of India. As an example, in addition to the budget of over 100 crore for the older IITs given by the GoI today as plan and non-plan funds, each IIT manages to get sponsored R&D projects worth Rs 20-50 crores every year. The end result of these projects is grenerally a report, few publications and one or two PhD degrees. There is not much correlation of the promised outcome with the outlays of these projects, in particular the massively funded so-called "Nationally Coordinated" projects. A clear conclusion is that money alone does not produce R&D in India. In sharp contrast, our own postgraduate students and PhDs are well known to be very productive in western academic institutions and infact are an important source of intellectual capital of USA. And, our own research students are also quite productive in terms of creating intellectual property when they work for research laboratories of several multinational companies located in India.

Why is our R&D in India not very productive ? The subject has been debated often enough by funding agencies and yet the situation has not changed much. The reason for the situation is clear enough (to those who conduct research) that our bureaucracy (both of the funding agencies and receiving institutions) has yet to create a proper mindset and to appreciate the need to function in a "business-like" manner.

Let us look at some of the weaknesses of our research funding system:

(a) There is no credible database of the names of the investigators and sponsored projects and the outcome of the projects in the country. The funding agencies do not bother to share information with each other, even in overlapping areas of R&D, let alone work together in any manner. As a result, it is not uncommon for an experienced faculty well versed with the "technology" of getting funded to get substantial funds for the same or similar project from more than one funding agency.

(b) The funding agencies have rarely made an effort to solicit or invite competitive proposals for nationally important projects.

(c) After going through the hassles of submitting a project and getting its approval which can easily take one or more years, the miserly release of funds in protracted phases by the agency and its poorly planned utilization presents horrifying, challenging and time consuming experience for any principal investigator.

(d) Most of the allocated funds are spent on importing equipment. But, very little money is provided for maintenance and accessories. Further, even simple chemicals and other materials have to be imported whereas these could be and should be produced in India if the funding agency pursues such a strategy diligently with a generous and liberal policy towards supporting entrepreneurs in academia.

(e) Only a few funding agencies conduct one or two reviews and assessment of the progress of the project through selected peers. Just how seriously this process is taken is obvious from the fact that very rarely has any project ever been cancelled for want of any visible outcome, or an investigator has been reprimanded or blacklisted.

The problems associated with all aspects of sponsored projects are many and must be recognized first and then solved if research has to become an attractive, effective and satisfying experience for any engineering faculty. A paradigm shift of a transparent, flexible, liberal, pragmatic, accountable and autonomous system is the need of the hour.

10.7 WHAT NEEDS TO BE DONE?

In order to understand what needs to be done, we need to analyze why US institutions are a magnet to post-graduate students from all over the world, including India. Recognizing that Knowledge is the driver of new economies world over, the policies of the US (as also several other advanced countries) government , the academic institutions as well as knowledge industries are periodically tuned to offer very attractive research ecologies to attract and retain some of the best talents of the rest of the world. Well- paying fellowships, excellent research facilities, stimulating environ, competitive, challenging and relevant research, and opportunity to work and to be absorbed by the industry are indeed very attractive incentives. If India wants its best graduates to do post-graduate studies and research in its premier institutions, it has to compete with other advanced countries in providing comparable facilities, environment, and opportunities.

Some measures for consideration by the Indian institutions are:

- Institutions must provide basic infrastructural facilities such as adequate electrical power and water which are lacking even in all IITs, our Institutes of National Importance, what to speak of universities and colleges in the country.
- Critical shortage of qualified and research oriented faculty which, besides undertaking R&D activities, will nurture the future generation of qualified faculty needs to be addressed to by all technical universities and premier engineering colleges by initiating post graduate and doctoral programmes and by a systematic and committed mentoring system in premier institutions.
- Sophisticated research facilities, whether centralized or available with individual researchers, should be made available to all legitimate users without barricades.
- Research students should be trained and allowed to use the facilities on their own for both experience and maturity.
- Research guidance should be allowed to only competent faculty members.
- Research students should have a choice of research topics and research supervisor(s) in any relevant departments of an institute.
- Research activities should be encouraged to be pursued seamlessly among different departments and faculties.
- Research students should be considered as research collaborators and respected for their independent views and approach.
- Due recognition should be given to the R&D contributions of the research students in terms of credits, awards and rewards.

- Research publications arising out of publicly supported projects and summary of MTech and PhD theses should be made available as an "Open Access" on an appropriate website, or a repository of the institution or any regulating agency designated for the purpose. This practice will stimulate transparency and competition, and prevent duplication as well as plagiarism.

- Research fellowships should be made financially attractive and dignified awards.

- Research students should be allowed to interact with industry and other institutions without bureaucratic tangles. They should also be given opportunity as tutors and teaching assistants and be paid for it.

- Faculty should enjoy full autonomy, with accountability, for bidding and for executing research projects in a hassle- free manner based on an honour code of conduct.

- Faculty should be allowed financial incentives, in addition to their salaries, from sponsored R&D and consultancy projects depending on time spent particularly during vacation periods. The concept of 9 month salary for the faculty and whatever additional earning possible during summer/winter vacation needs to be adopted.

- Faculty should get due recognition of R&D contributions in the form of promotions, awards, rewards, conference participation, special leave from teaching if required, etc.

- Institutional governance, procedures, rules and regulations for executing R&D projects, interactions with industry/other institutions should be made very flexible, liberal, pragmatic, and friendly. The academic institutions must usher in a business-like era of functioning.

- Funding agencies being primarily government are known notoriously for archaic bureaucratic procedures and for a sticky red tape. They must reform drastically to become facilitators with transparent, flexible, pragmatic, business-like procedures, rules and regulations.

- Competitive bids should be solicited/invited from competent R&D workers for important goal oriented projects which should be funded liberally and for adequately long durations.

- Hassle -free monitoring, auditing and grading of R&D projects by competent persons should be a rule rather than exception. The outcome of such projects should be publicized on an appropriate website of the funding agency for public scrutiny.

- Some academic institutions, in particular the IITs, have been acquiring intellectual property(IP) rights for inventions arising out of publically funded R&D projects and have also licensed the IP to private industry. However, there is a need for a formal and legalized process of empowerment of academic institutions as a not-for-profit society, along the lines of the well-known Bayh-Dole Act and the Stevenson-Wydler Act of USA, so that they are incentivised to actively negotiate the utilization and commercialization of their IP by sharing with, or by transferring to, the private industry.

10.8 ACADEMIA-INDUSTRY INTERACTION

Interaction between academia and industry in India in the form of joint or collaborative R&D projects has been rather insignificant in the past, partly because the confidence level between the two is very low and partly because the Indian industry, with some exceptions, is shortsighted and has short term goal of making profit by adopting or by buying whatever turn-key technology is available elsewhere. On its part, academia has also not been enthusiastic to develop interaction with industry even though it has been proven in the western world that stronger academic engagement with industry improves and strengthens basic research.

It must be recognized that the research profiles of industry and academia are expected to be vastly different and thus have little overlap. Academic research is characterized as : Scientific ; Open-ended; Long term; Unconstrained ; Knowledge generator ; Analytical; Peer reviewed; Degree requirement based; Human resource development oriented; Publication oriented; Constrained by financial and infrastructural resources. On the other hand, industrial research has its own identity which may be described as: Need based; Focussed, Short term; Real world problem; Professionally executed and managed; Viable product/process/application and profit oriented.

Just how academia and industry can meet to collaborate fruitfully requires special efforts. Collaboration between two or more partners requires a driving force derived from shared values, challenges or threats, and tangible benefits. Economic power is being increasingly determined by knowledge and innovation created and nurtured by knowledge institutions Paradigmatic shifts such as the internationalization of research and knowledge industries, globalized markets and competitiveness, outsourcing and globalized IPR protection have provided the required driving force for the interaction. These developments have called for a closer interaction and collaboration of knowledge institutions with industry, entrepreneurs, society and market. Indeed, the interaction has become essential for the survival and growth of both the academia and industry in a highly competitive world.

10.9 COLLABORATION AVENUES

Some suggested avenues for collaboration are:

- Establishing Chairs by the industry for industrial R&D, training, and school finshing life-long learning programmes.

- Periodic short or long term exchange visits of faculty and industry professionals.
- Appointment of Adjunct Faculty from industry by recognizing equivalence of the industrial experience with academic MTech and PhD degrees.
- Industry sponsored focussed and result oriented R&D projects.
- Industry sponsored post-graduate and research Fellowships in areas of interest to the industry.
- Short term training programmes for students in industry and for industry personnel in academia.
- Development of sandwich programmes for earning a degree by persons employed in the industry.
- Organisation of joint seminars, workshops, and special lectures.
- Sharing the use of expensive and sophisticated test and analytical facilities.
- Development of curricula and new topics for learning and research.
- Participation in evaluation of theses and research projects.
- Undertaking of joint sponsored R&D projects funded by government or private agencies.
- Undertaking joint consultancy projects.
- Participation in S&T parks, Entrepreneur development cells, Technology business incubators, Research consortia, Innovation Promotion Groups, etc.
- Establishing joint S&T based industrial enterprises with equity participation.
- Establishing new or converting existing facilities into dedicated industrial research centres.
- Participation in academic governance.
- Establishing institutional mechanisms such as technology foundation and venture capital.
- Synergising R&D activities to earn a brand name for both the industry and academia.
- Providing incentives and rewards for collaboration.
- Working together to integrate ethics, sustainability and social responsibility in engineering education as also in the conduct and pursuit of R&D.
- Initiating collaboration between industry, academia, professional institutions, and science and engineering academies to develop a database on a common website to provide information and to highlight the scope and achievements of sponsored R&D projects.
- Establishing a Think-Tank jointly with all stakeholders to assess and forecast emerging trends in engineering education, R&D and technologies.

10.10 STATUS OF COLLABORATION

- Industry sponsored R&D projects in academic institutions are not very common even today. However, spurred by the example of alumni of IIT Kharagpur setting up a research centre on telecommunications, alumni of other IITs and some private industries have come forward to set up a few research centres in each of the IITs primarily in electronics, communication and IT areas. But, serious and meaningful collaborative research in these centres is still not forthcoming as a matter of strategy for innovative growth.

- After persistent efforts, IIT Kharagpur was the first institution to receive several industry sponsored Chair professorships. Following the example, all IITs have now a large number of such Chair Professorships created by the alumni and industry. Unfortunately, these Chairs continue to be primarily for academic activities and have not yet been instrumental in creating any industry-academia collaborative R&D programme.

- During 90s, Planning Commission sponsored a very promising venture of Mission Mode Technology Projects to be undertaken jointly by the IITs and related industry. It took over two years of numerous meetings and presentations to get funds released partially from the government by which time much of the enthusiasm of collaborating industries vanished. This is an example of an excellent concept which got nipped in the bud, thanks to the Indian red tape. The Technology Development Board, TIFAC, etc are other agencies which are expected to nurture academia-industry interaction. The results so far are, however, not very inviting, particularly for small ventures by academics.

- Industrial consultancy is being increasingly appreciated as a challenge to both the faculty and the concerned industry. It is, however, confined to only few national institutions of repute and that too to a very few senior faculty members. Attractive incentives need to be provided to encourage the involvement of more faculty in industrial consultancy.

- The Department of Science and Technology (DST) has initiated several programmes to foster industrial product/process oriented R&D based on knowledge created by associated academic institutions. Presently taken care by its National Science and Technology Entrepreneur Board(NSTEB), there are 15 Science and Technology Entrepreneurship Parks (STEPs) in the country, compared to over 500 in China and over 50 in a tiny country of Israel. Only 5 of the STEPS in India are presently doing reasonably well and are nearly self supporting.

- Under the DST programme, IIT Kharagpur was the first IIT to set up a successful STEP on 100 acre land area donated by the state government. The institute liberalised rules of governance of the park and allowed selected faculty members to set up enterprises despite opposition and warnings from the central and state government. Several equity based joint ventures were also initiated with some industries. This model of allowing faculty to set up joint ventures along with their colleagues, students and industry on the basis of their research interests has now been adopted by all IITs and some public and private engineering institutions. All IITs have now established their own versions of STEP, Entrepreneurship Development Cells (EDC), or Technology Business Incubators (TBIU), etc. Also, over 150 EDCs / TBIUs have now been approved by NSTEB in various engineering colleges and universities. Besides the DST, several other government departments (such as MIT, DSIR, MHRD, UGC, AICTE) have now come forward to support such entrepreneurial activities in academic institutions. This is a healthy development. However, steps need to me taken to enhance the presently limited involvement of industry in such ventures.

- Challenging young persons to become Technology or Knowledge entrepreneurs to translate their dreams to action is a very important investment for any knowledge society. Though it is a far sighted experiment by DST and other agencies, its success so far is very limited. A lot of refining and tuning is needed, particularly in such directions as: (a) the mode of operation adopted by the funding agency and the associated institutions; (b) liberal hand holding of entrepreneurs by way of providing grants, venture capital and infrastructural facilities; and (c) assured transparency, ethical conduct, and protection and marketing of intellectual property.

10.11 SOME RECOMMENDATIONS

The following recommendations are meaningful only if we recognize that academics are one of the important pillars of the evolving Knowledge Society. And, to survive, compete and thrive in the tech-nomic globalization era, a Knowledge Society depends on creativity and innovations of the Knowledge disseminators and creators. The academic institutions, government S&T funding agencies and industries have a high stake in working together to usher in a new era of liberal, flexible and pragmatic governance and financial rules and regulations for conducting relevant and meaningful R&D in academic institutions.

- Collaboration between industry and academia be viewed as an asset and a future investment by all stakeholders.
- Major attitudinal and mindset changes (e.g; mutual trust, respect, motivation, spirit of cooperation) are required among all the stakeholders (faculty, students, research and support staff, industry).
- A regular and liberal budget needs to be provided by both the academic and industry partners.

- A visionary and intellectual crusader should be appointed to manage such collaborations.

- Both collaborating partners should set up a mechanism for involvement in the management and assessment of project and its outcome in a business-like manner.

- Financial incentives for collaboration, particularly to poorly paid faculty, need to be provided.

- Liberal sharing of intellectual and infrastructural resources among collaborators needs to be assured by suitable procedures.

- Protection and sharing of Intellectual Property is essential and needs to be assured.

- Strengthen technical knowledge system, source and innovation capabilities for collaboration.

- Provide liberal and flexible rules, and fast decision making process for R&D collaborations, recruitment, exchange visits, lectures, conferences, awards and rewards.

REFERENCES

1. India Vision 2020 – Towards a knowledge society.
2. National Knowledge Commission Reports to the Nation 2006 and 2007.
3. U.R. Rao, UG Review Committee of AICTE "Revitalising Technical Education", 2005.
4. Rangan Banerjee, Vinayak P. Muley, "Engineering Education in India" report on review sponsored by Observer Research Foundation, 2008.
5. R Natrajan, "Emerging Trends in Higher Education and Indian Scenario", 2002.
6. Pawan Agarwal "Higher Education in India: The need for change", 2006.
7. Gangan Prathap, B.M. Gupta, "Ranking of Indian Engineering and Technological Institutions for their research performance during 1998-2008, Current Science, 97, 304, 2009".

11

Technology Assisted Learning

"Professors will not be replaced by web pages; they will be replaced by Professors who know how to effectively use web pages."

— An unknown educationist

11.1 INTRODUCTION

Proper use of Information and Communication Technologies (ICT) can make quite a few problems in education simple. In the teaching-learning environments in the educational institutions, it is necessary to enhance the effectiveness of learning. With advancements in the communications and data management techniques, it is possible to disseminate and share information across the country. These advances need to be utilized to improve the efficiency of the system.

The Indian education system has marked imbalances in the access to quality infrastructure, content and process of learning. Computer and Networking technology is an important tool for enhancing the utilization of human resources. Government and policy makers have to focus on the promotion and development of information technologies. Society has to deploy technology for addressing the quality and relevance of learning and to develop adequately trained manpower to cater to the needs of the engineering profession.

11.2 STRENGTHENING COMPUTING, AND NETWORKING INFRASTRUCTURE

Computing and Networking infrastructure is essential for all technological institutions. This would enable them to make use of the advances in science and technology for enhancing the quality of education and to make them relevant to the rapidly changing socio-economic conditions. The existing Computing and Networking infrastructure in Central Institutions/ Universities is modest, at present. Attention should be focused on other government and private institutions. Improvements in the hardware/software/ campus networks and inter-connectivity between institutions across the country are essential.

In the premier institutions, each student has a computer. The scenario is different in most of the many other institutions (including government institutions), more than 10 students share a computer. To fully benefit from the new technologies in the learning environment, ideally each student should have a computer. As a strategy, new technologies that entails smaller, cheaper, and simpler platform could be adopted. As a starting point to bridge the gap not only amongst educational institutions but also in society, networked computing, at low prices should be made popular. A cheaper server-client based computing environment with open source software may be considered as a solution.

In order to handle the flow of information at high speed from external sources, all the institutes must improve the internal campus network. In most of the campuses, the application considered so far, is predominantly e-mail access and Web browsing. Since the scenario is changing, it is necessary to improve the networks in every campus. In addition, academic and laboratory areas in every college campus could have a WiFi network.

Reliable connectivity is essential for Technology Enhanced Learning. This includes, integration of Digital Library efforts of all institutions and integration of academic and research community across the higher education institutions. There is also a need to form a Computer Grid for High Performance Computing.

The focus has to be on better inter-connectivity between institutions and saving on international connectivity by consolidating the requirements of the institutions. A hybrid network of VSAT links has to be used. The network can be configured to have Voice-over IP facility to facilitate voice communication between the institutions. The department of space has to provide facilities for such connectivity.

11.3 NATIONAL PROGRAMME ON TECHNOLOGY ENHANCED LEARNING (NPTEL) AND NATIONWIDE NETWORK

The National Programme on Technology Enhanced Learning (NPTEL) was initiated by the Indian Institutes of Technology (IIT) and the Indian Institute of Science (IISc) Bangalore in 1999. The core curriculum development was later approved under the NPTEL programme and funded by the Ministry of Human Resource Development in 2003 for a budget of 205 million Indian Rupees for the creation of 120 Web-based courses, 120 video courses and encapsulation/ conversion of existing 110 video courses. Each Web-course comprises supplementary learning materials for 40 hours and video courses contain approximately 40 one-hour lectures per course. In addition, it was proposed that approximately 110 video-courses already recorded by IIT Delhi, IIT Madras and IIT Kharagpur—through their educational technology cells into digital video format (encapsulation)—be converted and be made freely available for the students and teachers of our country. The project has been completed and has ensured complete coverage of the first three years of undergraduate curriculum in five major engineering branches—Civil, Electrical, Electronics & Communication, Mechanical Engineering and Computer Science. These were supplemented by the courses in the core sciences and management.

Upgrading the standard of teaching is absolutely necessary, and is certainly a major task that needs to be accomplished. The country is struggling with the same obstacles as other countries in the world are—bridging the gap between "spoon feeding" in high- schools and self-reliant learning at the college/university. The teachers must not be viewed as the instructors feeding a set of information. The ethos of teaching is to provide optimal learning opportunities and to nurture creative and critical thinkers. Of course, one has to train the mind of the youngsters with appropriate rigor. Any pedagogical approach has to accept that the "training of mind" through proper track is necessary. Proper training helps every student to become self-reliant as soon as possible. The NPTEL material is quite a good model for providing the optimal learning environment.

The NPTEL (Phase-II) has already been approved by the MHRD and it going to start soon. The NPTEL-II is envisaged to develop course material for other Engineering and Science disciplines using the expertise available in the IIT system and IISc, as well as in partnership with well-known faculty in universities across the country. The course material will be in the form of video tapes, Web-based learning information as well as course-ware in the form of reference material. The huge reserve of well-known superannuated faculty members from the IIT system, IISc and other institutions across the country will also play a vital role in content creation.

The role of laboratory, experimental work and research is also a crucial element of Higher Education. Providing such training—when the system is very large—becomes difficult.

It has been found that technologies of virtual laboratory and simulation may help in fulfilling the requirements of training for the students in laboratory work. The experiments at one place are suggested to be perfected and also sharpened, especially in the light of new and complementary features of the knowledge available in open literature. It is recommended to fine tune the advanced features. The experts will conduct such thoroughly designed version of the experiments at one place and the procedure and data reduction will be transmitted to the remote classrooms at the Institutes where creating such experimental facilities are not possible due to the financial reasons.

11.4 DEVELOPMENT OF LEARNING MANAGEMENT SYSTEMS

Usually the Learning Management Systems (LMS) are web-based software to facilitate access to contents created by a group and the administration of content-distribution. LMSs are also used by industries (for example, financial services and biopharma) for training individuals. To be more precise, the **learning management system** is a tailormade software for delivering, tracking and managing courses for e-learning. The ICT can be leveraged in a big way through the efficient use of LMSs. The LMSs range from distributing courses over the Internet to offering online examinations. In many instances, acdemic institutions use LMSs for record-keeping as well as for the registration of the students and conducting quizzes. These can also be used for students' self-service (such as, registration, posting queries etc.), on-line assessment of assignments, collaborative learning through blogs, application sharing, group discussions etc. A very well known LMS "Moodle" provides such open-source community-based tools for e-learning. Moodle has a significant user base with about 2 million courses. Another indigenous LMS developed by IIT Kanpur is known as "Brihaspati". Development of such open-source LMSs should be funded and encouraged by the government.

11.5 VIRTUAL CENTRES FOR MANPOWER DEVELOPMENT IN EMERGING TECHNOLOGY AREAS

Admittedly, postgraduate education and the doctoral research in engineering and technology in the country are weak. Some departments or centres in some premier institutions and universities and a few CSIR laboratories are the exceptions. The major reason has been the failure to attract bright youngsters to these programmes. Students do not find the research environment in many of our institutions encouraging. Many programmes do not throw enough challenges to the intellectual abilities of the bright students. Not many institutions are working in cutting-edge areas of technology. Though the rate of scholarship/ assistantship has been enhanced recently, the effort is unable to stop the continuous outflow of bright people to the foreign countries.

We need to address these problems and to focus on postgraduate and research programmes in emerging technology areas. This can be done by setting up infrastructure on some high-potential technology areas. These shall be the collaborative initiatives of selected academic institutions, CSIR laboratories and the industry to create critical mass around the cutting-edge areas in technology. Collaboration shall be done over the network and communication links, with the provision for person-to person contacts, as and when needed. These initiatives can be named as the Virtual Centres of Excellence. Each of the initiatives would be adequately funded. The areas should be chosen carefully in order to generate manpower to improve the knowledge economy.

There has to be emphasis on sustainable and affordable technologies that are expected to benefit the masses. Focus could be also on disaster mitigation. The areas that suffer from adequate resources are: Synthetic Biology, Advanced Reactor (Nuclear) Technology, Energy (Hydrogen) Technology, Wireless Networking, MEMS Devices, Embedded Systems, Optomechatronics, Computer Security, Bio-informatics, and Computational Mechanics.

Under such initiatives, support should be provided for creating a collaborative learning and working environment over the Internet. Also, there must be some budget allocation for meetings and workshops.

11.6 RECOMMENDATIONS

With the advancement of technology, achieving integration and interoperability has become very important. In order to integrate e-learning services, content and applications effectively, the following requirements are to be met:

The e-learning framework must provide access to knowledge and related data anytime anywhere. The e-learning environment must provide a flexible workflow and process model that can be fine-tuned and configured to meet the need of the organization. The e-learning framework must allow additional components to be integrated easily using open software. The e-learning framework must allow content and other data to be exchanged and shared by tools and systems connected via internet. Through the course management tools (Learning Management Systems), syllabi and other education information are to be made more accessible to the students. There has to be some collaborative initiatives of selected academic institutions, CSIR laboratories and the industry to create critical mass around the cutting-edge areas in technology. Collaborations can be done over the network and communication links, with the provision for person-to person contacts, as and when needed.

REFERENCES

1. NPTEL Initiative available at *www.nptel.iitm.ac.in/pdf/NPTEL%20Document.pdf* Also browse [http://nptel.iitm.ac.in]
2. http://moodle.org
3. http://home.iitk.ac.in/~ynsingh/tool/brihaspati.shtml
4. Infosys campus connect program available at www.campusconnect.infosys.com
5. HRD in IT, A Document of MHRD, 2003

12

Innovation, Intellectual Property Rights, and Entrepreneurship

"Scientists investigate that which already is; Engineers create that which has never been."

— Albert Einstein

12.1 BACKGROUND

Innovation, intellectual property and entrepreneurship are closely linked to the engineering profession, since engineers improve existing engineering devices, services and systems(innovation), build completely new devices, artifacts and systems(intellectual property), and set up new business in construction, manufacturing or services (entrepreneurship). The engineering education system can play a major role in equipping graduates with skills of innovation and creativity, training them in identifying and protecting their intellectual property rights, and making them familiar with the requirements, characteristics and methodology for successful entrepreneurship.

12.2 INNOVATION

Innovation has been defined in literature in many ways, which include in general "the act of starting something for the first time, introducing something new" and specifically in technology as "an improvement to something already existing". The best definition on the Web appears to be: *Innovation is the conversion of knowledge and ideas into a benefit, which may be for commercial use or for the public good, the benefit may be new or improved product, processes or services.* Assuming some commonality in creativity and innovation, Sternberg's (1985, 1995) contention of the essential requirements for creativity is worth mentioning. According to Sternberg, a creative individual must possess intelligence (synthetic, analytic, and practical), knowledge of recognizing what is genuinely new, a thinking style which questions conventional wisdom instead of passively accepting such wisdom, and a non-conformist personality. He should enjoy his work (intrinsic motivation), be motivated either through financial or non-financial incentives including increased salary, promotion, commercialization of innovation, recognition and fame, and work in an environment which allows time and freedom to think, study and experiment without obstruction. (Standler 1998, 2002).

12.2.1 India's Heritage in Innovation

Indian society has a proud record at innovation. Indian music allows the artist to improvise and innovate as widely and creatively as he/she can within a restricted confine of a raga. Indian traditional craftsmen have shown growing innovative ability to produce object d'art on both familiar and unfamiliar themes. Indian temple architecture shows innovative fusion of many traditions, both Indian and western. Assimilation of foreign cultural influences brought about by Moghul and British rule and adapting them suitably and often innovatively to local conditions have been a prominent feature of the Indian ethos. In recent times, mechanical, electrical, chemical, and pharmaceutical small-scale industrial establishments have been able to adapt and innovate on foreign designs to meet local needs. India has also done a remarkable job of innovating and diffusing knowledge and technology, especially in agriculture. As a result of the "green revolution," India has transformed itself from a net importer to a net exporter of food grains. India's "white revolution" in the production of milk has helped it to achieve the twin goals of raising incomes of rural poor families and raising the nutrition status of the population. In the high-tech sectors of Atomic Energy, Space Technology, High Power Lasers, and Super Computer manufacture where international restrictions were imposed on import of materials, Indian scientists have demonstrated innovative approaches to develop technology for self-reliance against heavy odds. In the service sector, India has become a top global innovator during the last decade and is now one of the largest suppliers of innovative IT solutions. All this has been possible due to the Indian cultural heritage of philosophical enquiry, freedom of thought and expression, ability to meet challenges when forced, tolerance to alternative points of view, and a sense of national pride.

12.2.2 Role of Universities and Engineering Institutions in Promoting Innovation

Universities and Institutions of higher learning provide an ideal environment for promoting creative and innovative endeavors. Across the campuses, one encounters widespread prevalence of a spirit of enquiry among teachers and students, demonstration of individual freedom of thought through open academic debates, a reluctance to accept conventional wisdom without questioning which provides a fertile ground for testing new ideas and for experimenting with new processes or products. The University also offers easy availability of interdisciplinary assistance from colleagues and experts when needed, provides through its contacts with government, industry and community an awareness of societal and technological needs -both national and global, and fosters a co-operative research atmosphere inside the University for the creation and dissemination of new knowledge. The Engineering Institutions and University departments in Engineering prepare students for innovation through training in problem identification, analysis and solution, and in general research methodologies; develop among them and among the faculty a life-long learning attitude; and establish productive linkages with major exploiters of innovation e.g., the industry, and the government. Young students, with fresh and impressionable minds, spend those few years of their life in the University level Institutions when they are at the peak of their learning curve. They actively interact with their peers and their mentors and thus provide a very valuable and annually renewable resource for being groomed in becoming innovators and creators of new knowledge or developers of improved products, processes, or services. The University and Institutions promote innovation through student and faculty research, disseminate results through publications, conference participation, and public lectures, and often exploit innovation through commercialization of research results and patents, providing consultancy to industry, and setting up business incubators or technology parks. Both the Industry and the Government look up to the University and Engineering Institutions for receiving assistance in defining their long-term goals in economic and technological development, and in policy formulations for implementation. Apart from its acknowledged role of teaching, research and engagement with the community, and of providing an academic ambience for the creation and dissemination of new knowledge, the University is or can become a major contributor to the national wealth through intellectual capacity building and through exploiting the fruits of their own innovation and research.

12.2.3 Parameters for Success of Innovation Promotion and Dissemination

Although institutions of higher learning are ideal propagators and promoters of enquiry, research and critical evaluation, their success in developing innovators depends to a large extent on the attitude and mindsets of their students. Innovation and creativity have to be fostered in the student's attitude and learning methodology right from their childhood. The role of the school teacher both at the primary and secondary levels is the key to developing among the students an enquiring attitude and problem solving ability. Rote learning and accepting ideas without questioning at the school level can become a stumbling block for real creative and innovative work at the higher education level since students with such a background would not feel encouraged to traverse forbidden paths and look at newer ways of tackling problems. Roadblocks to success can come also if faculty in the colleges are not fully committed to imbibing new knowledge and advancing frontiers of knowledge, and the college management does not recognize and reward creative and innovative achievements. Effectiveness of university innovation promotion policy depends largely on free access to needed facilities—be it published literature, laboratory/field experimentation and peer interaction, on persistence and ability to learn from failures, on availability of critical but fair assessment of effort, on assurance for protection of intellectual property rights, and on a proper recognition and reward system.

12.2.4 Fostering a Culture of Innovation among Students and Faculty

Engineering Education offers enormous opportunities for promoting creativity and innovation since the processes of innovation are identical to those of problem solving and design which are the basic foci of all engineering education. Engineers often use the term "innovation" instead of creativity to describe the process that leads to insight or progress in a field, with a technique, or with a physical product. While innovation connotes a sense of inventing a thing as opposed to an idea or a theory, it is essentially a synonym for the creative process. Regardless of what one calls it, both innovation and creativity should lead one to the same end: to the exciting world of inventing and creating new knowledge, processes, and artifacts that push forward our science, technology, and art.

Stouffer, Russell and Oliva (2004) identify four basic steps in the creative process: 1) a notion or need (sensing, problem definition, and orientation); 2) an investigation of that notion or need (testing, preparation, incubation, analysis, and ideation); 3) an articulation of a new idea or solution (modifying, illumination, and synthesis); and 4) a validation process of that idea or solution resulting in an idea, theory, process, or physical product (communicating, verification, and evaluation). These four stages are identical to the normal engineering problem-solving and design process. Fostering a culture of innovation among students would require training in critical thinking, encouraging asking inconvenient questions in the classroom, thinking out of the box, looking at problems from multiple points of view, generating ideas and solutions including those which appear at first sight to be highly improbable, providing access to experimentation and other institutional facilities after normal working hours, frequent brainstorming among themselves with or without guidance of a teacher students be encouraged to work in groups, (preferably interdisciplinary) while tackling challenging problems, generate ideas and alternative solutions through guided brainstorming, refining their ideas list and presenting their solution to a larger class. Developing innovative and creative skill would need repetitive exposure to practice not left to one design project in the final semester .With some practice, students and teams of students will learn to set their imaginations free for generating multiple, original ideas and solutions. They will also learn to evaluate their solutions and reduce their list of possibilities in the light of project specifications and technical, economic and social constraints.

Problem-based/Project-based learning (PBL) has been proved instrumental in encouraging innovation and creativity. Challenges posed by institutional/national/international design competitions always bring out the best from senior students, and with some guidance from faculty they implement novel ideas and designs. Good institutions offer special innovation centers with adequate facilities for experimentation to motivated students either inside the Institution or in the Innovation/Science & Technology parks established by the Institution on its own or in collaboration with other agencies.

Richards (1998) recommends a series of activities to incite creativity when faced with an engineering problem:

- Immerse yourself in a domain or problem;
- Be prolific—generate lots of ideas;
- Use tools for representations and thoughts (e.g., brainstorming, notebooks, and sketches);
- Play with ideas;
- Avoid premature closure;
- Don't be afraid to be different;
- Be open and receptive to new ideas;
- Do it—practice your craft;
- Maintain a product orientation;
- Relax—indulge your diversions;
- Reflect—review what you have done;
- Have fun!

12.2.5 Teaching Creativity and Innovation as Part of the Curriculum

It is a common belief that creativity is an inherent gift that one either does or does not possess. A large body of research on the topic however suggests that creativity and the potential to innovate is inherent in most human beings. However, not all of us enjoy the environment that allows us to successfully translate our inherent creativity to innovative outputs. Torrence (1963/1977) considered as "Father of Creativity" spent his life-time in studying the nature of creativity and how it could be taught to students of all ages. Torrance's "Tests of Creative Thinking" effectively debunked the common assumption that IQ alone determined creativity. It also led to the now accepted belief that creative levels can be increased through practice and that creativity can be taught, effectively at all levels of education, from kindergarten children to graduate school students and researchers. Torrence's influence has spread to most institutions of higher learning in the United States and many courses in creativity and innovation are being offered at several institutions in many undergraduate and post-graduate programs. Some engineering professors in U.S. make creativity an explicit component of their courses (Ghosh 1993; Masi 1989; Richards 1998). Their effort is financially supported in many cases by the National Collegiate Inventors and Innovators Alliance (NCIIA) which has a program to "foster invention, innovation, and entrepreneurship in higher education as a way of creating innovative, commercially viable, and socially beneficial businesses and employment opportunities in the United States. The program was founded on the premise that invention, innovation, and entrepreneurship are essential components of the higher education curriculum and vital to the nation's economic future. The NCIIA works with colleges and universities to build collaborative experiential learning programs that help nurture a new generation of innovators and entrepreneurs with strong technical and business skills and the tools and intention to make the world a better place". The NCIIA is supported by the Lemelson Foundation which was founded by one of the most prolific U.S. inventors, Jerome Lemelson, and his wife Dorothy in 1993. Lemelson holds over 575 U.S. patents for inventions ranging from key components for the fax machine, VCR and cordless telephone to automated manufacturing systems.

Concerned with the decline in the United States' industrial productivity, Lemelson also developed his E-Team concept, where the "E" denotes both Excellence and Entrepreneurship. In his interview for the "Inventors" series on the Discovery Channel, he stated:*" What I consider to be one of my best innovations...an E-Team is a group of students who train to go into business and develop products in the future while at school."*An E-Team is a group of students from engineering and other disciplines, along with some faculty and professional mentors who pursue the development of an idea or invention with the desired result to be the licensing of new products, technologies or the startup of entrepreneurial ventures. Since 1994, the multidisciplinary E-Team concept has been adopted at several US universities with financial support from the NCIIA: Nevada, Lehigh, Virginia, Rowan, Illinois Institute of Technology, Rensselaer Polytechnic Institute, and Swarthmore College among others. Several formats have been used which include a capstone course at the final year undergraduate level, projects at the innovation and entrepreneurship center, part of an engineering clinic, and apprenticeship for building prototypes, writing patents and developing business plans. It would be desirable for Indian Engineering Institutions to set up Innovation centers on the lines of "The Lemelson Center for Invention, Innovation and Entrepreneurship at the University of Nevada, Reno", include creativity, innovation and entrepreneurship training as part of the engineering curriculum, and associate themselves with Science and

Technology Parks to help students develop their innovative skills and implement their ideas. AICTE could also develop and establish a consortium of Engineering institutions devoted solely to encouraging, fostering and supporting innovative ideas generated by students and faculty and converting them to commercial products with the support of interested industry or through venture funds.

12.3 INTELLECTUAL PROPERTY RIGHTS

12.3.1 Definition and Concept of Intellectual Property

Intellectual property refers to creations of the mind: inventions, literary and artistic works, and symbols, names, images, and designs used in commerce. Intellectual property is divided into two categories: Industrial property, which includes inventions (patents), trademarks, industrial designs, and geographic indications of source; and copyright, which includes literary and artistic works such as novels, poems and plays, films, musical works, artistic works such as drawings, paintings, photographs and sculptures, and architectural designs.

12.3.2 Rights Relating to Intellectual Property

The Universal Declaration of Human Rights adopted and proclaimed by the UN General Assembly on December 10, 1948 recognizes "the right to the protection of the moral and material interests resulting from any scientific, literary or artistic production of which he is the author" (Article 27-2), balanced by "the right…to share in scientific advancement and its benefits" (Article 27.1). There is some conflict between the public interest in accessing new knowledge and the products of new knowledge, with the public interest in stimulating invention and creation which produces the new knowledge and products on which material and cultural progress may depend. The conflict is usually minimized through limiting the exclusive right of the inventor or author to a limited period of time after which the knowledge becomes public property.

Intellectual property rights (IPR) are thus perceived as a bundle of exclusive rights over creations of the mind, both artistic and commercial for a given period of time . Artistic creations are covered by copyright laws, which protect creative works, such as books, movies, music, paintings, photographs, and software, which give exclusive right to the copyright holder for controlling the reproduction or adaptation of such works for a certain period of time. Commercial innovations are covered under "industrial properties", as they are created and mostly used for industrial or commercial purposes. A patent may be granted for a new, useful, and non-obvious invention or new innovation and gives the patent holder a right to prevent others from practicing the invention/innovation without a license from the inventor for a certain period of time. A trademark is a distinctive sign which is used to prevent confusion among products in the marketplace. An *industrial design right* protects the form of appearance, style or design of an industrial object from infringement. A trade secret is an item of non-public information concerning the commercial practices or proprietary knowledge of a business. Public disclosure of trade secrets may sometimes be illegal.

The term *intellectual property* denotes the specific legal rights described above, and not the intellectual work itself. By giving exclusive rights to their creations/inventions/innovations, the author or inventor is provided with an incentive for developing and sharing the information rather than keeping it secret. The legal protections granted by IP laws are credited with significant contributions toward economic growth.

12.3.3 Protection of Intellectual Property Rights

Most countries have enacted laws to protect intellectual property for two main reasons. One is to give statutory expression to the moral and economic rights of creators in their creations and the rights of the public in access to those creations and the second is to promote, as a deliberate act of Government policy, creativity and the dissemination and application of its results and to encourage fair trading which would contribute to economic and social development. Copyright and Patent laws have been on the Statute books of most countries, although strictness of the enforcement of these laws vary from country to country. While most developed nations have elaborate mechanisms to detect and punish infringements, many developing countries lack the capacity for proper detection of infringement of the laws and awarding appropriate punishment.

Recent studies have shown that industries which rely on IP protections are estimated to produce 72 percent more value per added employee than non-IP industries. (Economic Effects of Intellectual Property-Intensive Manufacturing in the United States, Robert Shapiro and Nam Pham, July 2007). Further, a joint research project of the WIPO and the United Nations University measuring the impact of IP systems on six Asian countries found "a positive correlation between the strengthening of the IP system and subsequent economic growth. (Measuring the Economic Impact of IP Systems, WIPO, 1997). The Convention Establishing the World Intellectual Property Organization (WIPO), concluded in Stockholm on July 14, 1967 (Article 2(viii)) provides that "intellectual property shall include rights relating to: literary, artistic and scientific work, performances of performing artists, phonograms and broadcasts, inventions in all fields of human endeavor, scientific discoveries, industrial designs, trademarks, service marks and commercial names and designations, protection against unfair competition, and all other rights resulting from intellectual activity in the industrial, scientific, literary or artistic fields." Several other International agreements have been reached by many countries for protecting different aspects of Intellectual Property, some of the important ones have been revised and amended from time to time to suit changing environment or inclusion of specific items due to advances in technology. They include the following:

(a) *The Paris Convention for the Protection of Industrial Property* of March 20, 1883, as revised at Brussels on December 14, 1900, at Washington on June 2, 1911, at The Hague on November 6, 1925, at London on June 2, 1934, at Lisbon on October 31, 1958, and at Stockholm on July 14, 1967, and as amended on September 28, 1979; it includes "the repression of unfair competition" among the areas of "the protection of industrial property"; and states that "any act of competition contrary to honest practices in industrial and commercial matters constitutes an act of unfair competition" (Article 10*bis*(2);

(b) *The Berne Convention for the Protection of Literary and Artistic Works* of September 9, 1886, completed at Paris on May 4, 1896, revised at Berlin on November 13, 1908, completed at Berne on March 20, 1914, revised at Rome on June 2, 1928, at Brussels on June 26, 1948, at Stockholm on July 14, 1967, and at Paris on July 24, 1971, and amended on September 28, 1979;

(c) *The Rome Convention for the Protection of Performers, Producers of Phonograms and Broadcasting Organizations* as accepted by members of the World Intellectual Property Organization on October 26, 1961, this agreement extended copyright protection for the first time from the author of a work to the creators and owners of particular physical manifestations of intellectual property, such as audiocassettes or DVDs; and

(d) *The Treaty on Intellectual Property in Respect of Integrated Circuits* (Washington, 26 May 1989.

12.3.4 Trade Related Intellectual Property Rights (TRIPs)

The Agreement on Trade Related Aspects of Intellectual Property Rights (TRIPS) is an international agreement administered by the World Trade Organization (WTO) that sets down minimum standards for many forms of intellectual property (IP) regulation. It was negotiated at the end of the Uruguay Round of the General Agreement on Tariffs and Trade (GATT) in 1994. Specifically, TRIPS contains requirements that nations' laws must meet for: copyright rights, including the rights of performers, producers of sound recordings and broadcasting organizations; geographical indications, including appellations of origin; industrial designs; integrated circuit layout-designs; patents; monopolies for the developers of new plant varieties; trademarks; trade dress; and undisclosed or confidential information. TRIPS also specifies enforcement procedures, remedies, and dispute resolution procedures. Protection and enforcement of all intellectual property rights shall meet the objectives to contribute to the promotion of technological innovation and to the transfer and dissemination of technology, to the mutual advantage of producers and users of technological knowledge and in a manner conducive to social and economic welfare, and to a balance of rights and obligations. Because ratification of TRIPS is a compulsory requirement of World Trade Organization membership, any country seeking to obtain easy access to the numerous international markets opened by the World Trade Organization must enact the strict intellectual property laws mandated by TRIPS. For this reason, TRIPS is the most important multilateral instrument for the globalization of intellectual property laws.

The TRIPS agreement introduced intellectual property law into the international trading system for the first time and remains the most comprehensive international agreement on intellectual property to date. In 2001, developing countries, concerned that developed countries were insisting on an overly narrow reading of TRIPS, initiated a round of talks that resulted in the Doha Declaration. The Doha declaration is a WTO statement that clarifies the scope of TRIPS, stating for example that TRIPS can and should be interpreted in light of the goal "to promote access *to medicines for all.*

12.3.5 IPR Protection in India

There had been some concerns in the recent past about the protection of Intellectual Property Rights in India. The Embassy of India in Washington had issued a policy statement back in 2002 to allay such concerns. The relevant portions of this statement are summarized here: 'There is a well-established statutory, administrative and judicial framework to safeguard intellectual property rights in India, whether they relate to patents, trademarks, copyright or industrial designs. Well-known international trademarks have been protected in India even when they were not registered in India. The Indian Trademarks Law has been extended through court decisions to service marks in addition to trade marks for goods. Computer software companies have successfully curtailed piracy through court orders. Computer databases have been protected. The courts, under the doctrine of breach of confidentiality, accorded an extensive protection of trade secrets. Right to privacy, which is not protected even in some developed countries, has been recognized in India".

India is a signatory of TRIPs in the Uruguay Round agreement of 1995. It has amended her existing laws in order to make it TRIPs-compliant. In the last few years, India has enacted fully TRIPs-compliant Trademarks Act, Copyright Act, Designs Registration Act, Geographical Indications Act and Protection of Layouts for Integrated Circuits Act. A novel Plant Varieties Protection and Farmers Rights Act 2001 and the Bio-diversity Act 2002 are also in place.

Intellectual property in India is administered by five central Ministries in terms of the Allocation of Business Rules of the Government. Industrial property which includes patents, trademarks, industrial designs and geographical indications is regulated by the Department of Industrial Policy and Promotion of the Ministry of Commerce & Industry. Copyrights are looked after by the Ministry of Human Resource Development. Legislation on Plant Varieties and Farmers' Rights Protection is handled by the Ministry of Agriculture. The Ministry of Information Technology is responsible for implementation of the Information Technology Act and the Semiconductor IC Layout Designs Act. Implementation of the Biological Diversity Act is the responsibility of the Ministry of Environment and Forests. Besides these administrative Ministries, there are a number of other Ministries and Departments such as Information & Broadcasting, Tribal Affairs, Culture, Ministry of Micro, Small & Medium Industries which are also involved with either enforcement or commercialization of IP. As on date, India is fully in compliance with its international obligations under the TRIPs Agreement.

12.3.6 Training Engineering Students in Protection of Intellectual Property Rights

Law Schools have been quick in introducing Intellectual Property Rights and their protection in the MBA curriculum in most parts of the World including India {*IIT Kharagpur Rajiv Gandhi School of Intellectual Property Law-Three year LLB programme with specialization in Intellectual Property and One and Half Year Post Graduate Diploma in Intellectual Property Law (PGDIPL) (*www.iitkgp.ac.in*), WIPO-IGNOU One Year Post Graduate Masters in IP Diploma through distance education (*http://www.ignou.ac.in/schools/sos/pgdipr1.htm#wipo*), One Year PG Diploma in IPR Law through distance education – National Law School of India University Bangalore: (*http://www.nls.ac.in*)}*. Engineering schools have been slow in this aspect although scientists and engineers are the most prominent contributors to industrial intellectual property through creation of new knowledge, inventions and innovations, new industrial designs, new software solutions and new patentable products, processes and systems.

In the United Kingdom, with the support of the UK Higher Education Academy, Ruth Soetendorp of the Bournemouth University (UK) was appointed as Team Leader for a project on "Intellectual property for engineers: a curriculum development project" which started functioning in 2005. The project team is a unique blend of engineering and law academics including Professor Jim Roach at Bournemouth's School of Design, Engineering & Computing, Bill Childs and Dr. Rob McLaughlan of the Faculties of Law and Engineering at the University of Technology, Sydney, respectively. As well as receiving funding from Engineering Subject Centre, the project has been awarded funding from the Higher Education Academy's Law Subject Centre highlighting its interdisciplinary focus. Earlier in 2002, Ruth had presented her paper 'Intellectual property rights awareness: a business enterprise skill for engineering education' at the Australasian Association for Engineering Education 13th Annual Conference, held in Canberra and MacLaughlan was one of the co-authors of a paper on "Engineering enterprise through IP education: what is needed?" in the Proceedings of the 2005 ASEE/ AaeE 4th Global Colloquium on Engineering Education. The project is a major initiative in bringing the awareness and need for the education and training of engineering students in IPR protection. In United States and Europe, research universities and major Engineering Institutions give regular exposure to IPR and patenting procedures to their undergraduate and graduate students involved in research and are encouraged to develop business plans for their research output for commercial exploitation. Special courses are also available from Business schools as an elective in their engineering curricula.

In India, MHRD has taken up the initiative in setting up six general IPR chairs in the Universities of Allahabad, Delhi, Pune, and Madras, and at the National Law School of India University at Bangalore and Cochin University of Science and Technology; ten new chairs in specific areas: three on IP management at IIMs Ahmedabad, Kolkata and Bangalore, five in the areas of IPR relating to patents, trademarks, industrial designs and geographical indicators at IITs Delhi, Kharagpur, Madras, Bombay and Kanpur, and two on IPR and development at Delhi School of Economics and JNU center for economic studies and planning.

In Engineering Education, IIT Delhi has taken a major lead. As early as 1992, it sensitized its own Faculty in the area of IPR through a workshop, followed up by another workshop in 1993 for faculty of all other IITs and IISc, and two workshops in 1994 jointly with CII and WIPO for students and SMEs. Since then, several workshops and seminars have been held from time to time an International Conference on Intellectual Property education and Training was also organized in 2001 in collaboration with MHRD and WIPO and a faculty development program on Intellectual Property Rights (for faculty members of engineering colleges) was held as recently in March 2008 under AICTE/QIP sponsorship. Department of Management Studies of the institute developed a full semester course (SM802: Management of IPRs) in 1997. This course is being offered regularly as an elective to all Post-graduate and III/IV year B.Tech. students of the institute. Timely focus on IPRs has resulted in an increasing number of patent applications filed by the Institute. Many other Institutes have followed the IIT Delhi lead and established IPR cells and given exposure to engineering undergraduate and post-graduate students. It is recommended that all engineering institutions have an IPR promotion cell and introduce student to evaluate their research for eligibility for patenting and commercial exploitation.

12.4 TECHNO-PRENEURSHIP

12.4.1 Introduction

The process of developing and fostering proof-of-the concept, and/or building an old business/undertaking out of innovative ideas is termed "entrepreneurship".

A techno-preneur is an entrepreneur who creates and innovates on the basis of scientific and engineering knowledge to build something of recognized value for society around perceived opportunities.

It is important to keep in mind that entrepreneurship is not a job but a spirit of creativity, innovation, adventure, challenge, excitement, and achievement. It is akin to an adventurer who climbs Mt Everest "because it is there" even though knowing fully well that he/she may not come back alive. An entrepreneur has a special DNA with such qualities as:

- Self-esteem as a driving force.
- Gut-feeler, instinctive and speedy.
- Multidimensional in approach.
- Knowledge-Inspired.
- Dreamer and Visionary with a broad mission.
- Self Life-coach who recovers from a fall by himself.
- Life-long learner who believes that knowledge is never wasted.
- Adaptable to changes in systems and persons and the one who makes a significant difference to whatever he does.
- Creative and innovative.
- Spots and exploits opportunities.
- Finds resources and competencies required to exploit opportunities.
- Good team-builder and networker.
- Faces adversity and competition professionally.
- Manages change and risk.

12.4.2 Why Entrepreneurship?

- The old industrial economy has been run by the power of Money, Manpower, Machinery, Materials and Methods – the so-called 5M's. The new emerging economy of the present century is being largely determined by the Mind power, the 6th M. In this changing scenario, only the most adaptable to mind power will survive.
- As a 'result of globalization of science, engineering and manufacturing technologies, trade and markets, business practices and services, a significant shift of emphasis from a decaying industrial economy composed of large firms to a knowledge-based entrepreneurial economy driven by knowledge based innovative technologies has taken place. Rapid changes in technology and innovations have lead to shorter product cycle, and strong emphasize on quality, environment standards, sustainability, IPR, and other issues. With the availability of various financing options, entrepreneurial activity is gaining momentum worldwide. Higher education institutions are becoming an important link in the innovation cycle and in promotion of entrepreneurial instruments.

- Growth of global economies and life styles of civilized societies are being increasingly determined by knowledge and innovation created and nurtured by knowledge institutions. The demand for a closer interaction of such institutions with entrepreneurs, communities and industry is increasingly becoming louder.

- Entrepreneurship strengthens the knowledge system, converts knowledge to intellectual property, promotes techno-ventures for commercialization of technologies, creates wealth and enhances technology competitiveness and the tech-image of the country. Further, it creates new business and creative opportunities, jobs and services, and thus promotes regional as well as local development.

- The shining example of the one trillion dollar economy created and nurtured by the Innovative Machines of the Silicon Valleys of the world are a testimony to the spirit of entrepreneurship which are well documented by the World Bank reports. Entrepreneurs in Silicon Valley, USA create literally new firms and new millionaires every week.

12.4.3 Mechanisms to Nurture Innovation and Techno-preneurship

Recognizing the important role of entrepreneurs in the economic growth, the concept of Science & Technology Parks and Incubators for techno-preneurs originated in USA during 1950s. Today, USA has over 1000, China has over 500, Germany has over 200 such platforms. The concept originated in India during 1980s . Presently, India has 13 S&T Entrepreneurship (STEP) parks, and some 50 Incubators of different types which have been funded by the National S&T Entrepreneuship Development Board (NSTEDB) of the Department of S&T (DST), Government of India. More recently, Department of Biotechnology (DBT), Ministry of Information and Technology (MIT), and Ministry of HRD have taken keen interest in sponsoring and supporting similar platforms for entrepreneurship in specialized areas.

Whereas IIT, Kharagpur has set up a 100 acre area STEP, other IITs have suitable platforms akin to Incubators. With a very few exceptions, most of the 2000 private engineering colleges and technical universities and university departments have not yet started any entrepreneurship activity.

The variety of platforms created all over the world to nurture entrepreneurship include:

Research Parks, Technology Parks, Science & Technology Entrepreneurship Parks (STEP), Entrepreneurship Development Cells (EDCs), Industrial Parks, Innovation Centres, Technology Incubators (TIs), Technology Business Incubators(TBIs), Business Incubators, and Specialized Parks (such as Biotech, Information Technology, Electronics Parks).

Following is a brief description of the functions of some of these entities.

- **Research Park** provides a physical environment for interaction between academics, researchers, commercial organisations and entrepreneurs to advance scientific and technological knowledge. Manufacturing except generation of prototypes is prohibited here.
- **Science Park** provides affiliations with academia and research institutions with emphasis on development work, prototype production and technology transfer.
- **Technology Park** provides proximity to university, research institutes, applied research and manufacturing focus, and commercial application of advanced technologies.
- **Business Incubator** is an economic development tool for occupants not limited to technopreneurs. It provides shared office services and specialist services which help improved growth rate with reduced failure rate.
- **Technology Business Incubator (TBI)** is a low cost facility to incubate technological ideas or develop technologies of products /processes for commercialization in the market place. It helps young firms to survive and grow faster by providing them with specialized technical, financial, management and marketing support services during the critical period of the start-up phase of a business venture .

Depending on the type of activities, Incubators are further classified as: University Incubator having incubatees primarily from the faculty, students and alumni; Regional or Rural Incubator, having incubatees who are interested only in rural-based relevent technologies; Single Business Incubator which houses incubatees with one common business interest; and Virtual and Semi-Virtual Incubator which do not provide in-house services to the invitees but allow access to mentoring, advisory and consultancy processes.

Whatever be the nomenclature, all special purpose vehicles have a common core objective which is to nurture entrepreneurship in collaboration with institutions of higher learning, research and development. More successful ones have flexible and pragmatic policies with a broader perspective and are just called incubators or TBIs which allow an incubatee to move seamlessly from a concept to pilot scale production and commercialization in the same place before moving out of the incubator within a reasonable period.

12.4.4 Role of Government and Academic Institutions

Despite the efforts of the NSTEDB, entrepreneurial activities in India are still in an infancy stage. Some attractive incentives are required to accelerate the process so that most technical academic institutions in the country would have an active incubation centre. Both the government and the academic institutions must recognize the importance of this sort of activity and adopt significant measures to activate and popularize it.

Some suggestions for a desirable government role are:

1. Close interaction of academic institutions with entrepreneurs should be incentivised and mandated as a societal responsibility on the part of academia
2. Entrepreneurial activity be recognised at par with other formal teaching-learning processes and R&D activities which is worthy of recognition in terms of a suitable diploma or certificate from the mentoring academic institution.
3. Faculty members and students should be encouraged to participate in entrepreneurial activities as advisors, consultants, or part-time partners as long as their primary academic responsibilities are not compromised.
4. To be viewed as an investment in the economic future of the country, the government should fund entrepreneurial activities liberally like R&D projects on a long range basis and also provide liberal grants, fellowships, and awards to entrepreneurs.
5. Academic institutions should be allowed to create a venture capital out of their savings and consultancy earnings. They may be allowed to invest the venture capital in incubated businesses on equity basis or joint ventures. The academic institutions should be given the status of not-for-profit Societies so that any profit is ploughed back into the institution in whatever form it is needed.
6. Entrepreneures should be given the incentive of income tax benefits during incubation and pilot plant production stage.

12.4.5 Role of Academic Institutions

1. Academic institutions need to bring about attitudinal and mindset changes among all the stakeholders namely, faculty, students, alumni and support staff to accept entrepreneurship as an important academic and R&D activity of the institution.
2. Both faculty and students should be allowed to participate in entrepreneurial activities subject to meeting other academic responsibilities.
3. A sense of belonging among the techno-preneurs must be developed by providing teaching–learning, R&D facilities and part time registration for post graduate degrees/diplomas for upgrading their knowledge without barriers.
4. The principle of growth and progress should be based on some sort of equity partnership, or creation of a joint venture.
5. An appropriate technical knowledge system and business development services should be made available by the institution.
6. The institutions should adopt simple, liberal, pragmatic and dynamically flexible rules of business/governance for entrepreneurial activities.

7. It is essential to hold hands of entrepreneurs by liberally sharing intellectual and infrastructural resources.

8. A legal system for providing protection of Intellectual Property Rights must be in place.

9. Support for international cooperation and technological transactions in the international market should be provided, if needed.

10. Assistance need to be provided for obtaining financial support from sponsored projects, consultancy, and venture capital from the institution and/or outside agencies

REFERENCES

1. Sternberg, R.J. (1985): Beyond IQ: A Triarchic Theory of Human Intelligence, New York, Cambridge University Press.

2. Sternberg, R.J. and Lubart, T.I (1995): Defying the crowd: Cultivating Creativity in a Culture of Conformity, The Free Press, 1995.

3. Standler (1998,2002): Creativity in Science and Engineering by R.B. Standler at http://www.rbso.com/create.htm copyright 1998,updated May 2002.

4. Stouffer, W. B., Russell, J. S. and Olivia, M.G., "Making the Strange Familiar: Creativity and the Future of Engineering Education", *Proceedings of the 2004 American Society for Engineering Education Annual Conference & Exposition*, June 20-23, Salt Lake City, USA.

5. Richards, L.G. (1998). "Stimulating creativity: Teaching engineers to be innovators," Proceedings of 1998 IEEE Frontiers in Education Conference, 3, 1034-1039.

6. Ghosh, S. (1993). "Exercise in inducing creativity in undergraduate engineering students through challenging examinations and open-ended design problems," IEEE Transactions on Education, 36(1) 113-119.

7. Masi, J.V. (1989). "Teaching the process of creativity in the engineering classroom," Proceedings of 1989 IEEE Frontiers in Education Conference, 288-292.

8. Torrance, E. P. (1963). Creativity, National Education Association. Washington, D.C.

9. Torrance, E. P. (1977). Creativity in the Classroom, National Education Association. Washington, D.C.

10. National Collegiate Inventors and Innovators Alliance: Activities available at www.nciia.org/

11. The Lemelson Foundation; details available at www.lemelson.org/

12. Robert Shapiro and Nam Pham: Economic Effects of Intellectual Property-Intensive Manufacturing in the United States, July 2007. available at en. wikipedia.org/wiki/Intellectual_property.

13. Measuring the Economic Impact of IP Systems, WIPO, 1997 available at http://www.wipo.int/portal/en/news/2007/article_0032.html.

14. The Doha declaration available at en.wikipedia.org/wiki/Doha_Declaration,also at www.wto.org/english/tratop_e/dda_e/dohaexplained.

13

National Initiatives

"Just as the race to the moon galvanized an earlier generation, the education race must mobilize us to meet the demands of the 21st Century."

— Jon Corzine

13.1 BACKGROUND

There has been growing national concern on the deteriorating standards of technical education in the country, and on the need for reducing the shortage of qualified teachers. Major manufacturing and service Industries are also demanding a quantum jump in research facilities for fostering a research culture in most engineering Institutions, so that they can produce manpower with very high levels of creative and innovative abilities and thus assist Indian Industry in increasing the competitiveness of their products and services. Many Initiatives have been taken in the last few years both by the Government, and the Private sector, some of which are enumerated in the following section.

13.2 RECENT INITIATIVES FOR IMPROVING ENGINEERING EDUCATION

13.2.1 Government Initiatives

The Government at the Central level has through consultation with experts produced a large number of Policy papers, and has implemented many of them through new policy initiatives. Most of the documents, details and statements listed below have direct linkages with and impact on Engineering Education and are available on the Web (See References).

- **National Policy on Education (1986/92/2000):** Revised at intervals lays down policy guidelines on all sectors of education.
- **Establishment of NAAC, NBA(1994):** For setting up accreditation mechanism for quality assurance in Higher Education(NAAC) and in Technical education (NBA).
- **Technology Vision of India 2020 (1996):** A futuristic document prepared after exhaustive consultation among scientists and finally accepted by the

Cabinet, outlines a roadmap for the country becoming developed by 2020.

- **Information Technology Action Plan (1998):** Basic policy document and Roadmap for assisting IT Industry which has helped the growth of IT industry in India.

- **Encouraging Private Investment in Professional Education (Since 1980s):** Efforts resulted in increasing availability of engineering education many fold.

- **Liberal Grant of Autonomy-Deemed University Status, IIITs, NITs (1994-2008 and continuing):** For removing the shackles of government and university control on institutional initiatives for improving curricula, evaluation mechanisms and general standards of education.

- **Ambani Report** on "A Policy Framework for Reforms in Education" submitted to the then Prime Minister as part of a special subject group on Policy frame work for private investment in Education, Health and Rural Development (April 2000).

- **New Science and Technology Policy (2003):** Revision of original plan to accommodate new global realities.

- **Setting up and Educational Satellite (2003):** For disseminating educational material and networking among institutions, using the Doordarshan Gyan Bharti Channel and permitting use by other institutions for beaming their distance education and other programs.

- **Transforming India into a Knowledge Superpower (2003):** For Setting up guidelines for the rapid growth of knowledge and knowledge based industry.

- **Upgrading Technical Education System-Tech Ed.III, and TEQIP (2003-2008):** World Bank assisted projects for improving quality of polytechnic and engineering education and promoting research.

- **Setting up Indian Institutes of Information Technology, Design and Manufacturing (IIITDM) at Kancheepuram and Jabalpur in 2003-04:** These institutes are to provide a sustainable competitive advantage to the Indian industry in the area of design and manufacturing of new products.

- **Draft National Biotechnology Plan (2004/05):** For Initiating Discussion on an emerging sector.

- **Setting up of the National Knowledge Commission (2005):** The National Knowledge Commission is a high-level advisory body to the Prime Minister of India, with the objective of transforming India into a knowledge society. It covers sectors ranging from education to e-governance in the *five focus areas* of the knowledge paradigm: Access to knowledge, knowledge concepts, creation of knowledge, knowledge application and delivery of services. A major initiative which includes policy directions for improving quality of educational services, details of important recommendations in Section 12.2

- **Finance Minister's allotment of an additional INR 1000 million to Indian Institute of Science, Bangalore** to become a world level university (Budget speech 2005).

- **Prime Minister** speech inviting active public –private partnership while dedicating the Bharti School of Telecommunications Technology and Management at IIT Delhi, and promising liberalizing rules and procedures to make the partnership effective (March 20, 2006).

- **Finance Minister's** allotment of an additional INR1000 million each to Universities of Mumbai, Kolkata, Chennai and the Punjab Agricultural University to make them world class (Budget Speech 2006).

- **Moily Committee Report (2006):** Develops a strategy for implementation of OBC reservation in Central Institutions.

- Establishment of Indian Institute of Science, Education and Research(IISER) at Pune and Kolkata (2006-07).

- Proposals for Large Scale expansion of Higher Education and Research in the 11th Plan including establishment of new IITs, IIMs, IISERs, IIITs and World class central Universities which would have constituent engineering departments/institutions (2007-2012); details in section 12.3.

13.2.2 Private Sector Initiatives

The Private Sector Institutions and Industry also have taken many initiatives for improving the quality of engineering education and research in the country, some important ones are listed below:

- **Establishment of Indo–US Collaboration in Engineering Education (IUCEE 2007):** The goal of the Indo US Collaboration for Engineering Education is to improve the quality and global relevance of engineering graduates in the US and in India by scaling up promising collaborations between the two countries. The IUCEE will focus on increasing the number of engineering faculty in the US and in India who collaborate on research and teaching and who will be able to better prepare engineers for the global economy. The collaboration has so far held workshops and teacher competence up gradation programs both in India and the U.S. and is in the process of setting up permanent centers for teacher training through Indo-US faculty Leadership Institutes. It has collaboration with ASEE, World Bank Institute, Deshpande Foundation, Pan-IIT Forum and several Indian and US Institutions and Industry.

- **Amrita University-US Collaboration (December 2005):** In a landmark agreement, that was inked during the visit of the Prime Minister to **Washington DC on July 20th,** Gurupurnima day, five leading American Universities joined Indian institutions led by AMRITA Vishwa Vidyapeetham, ISRO, and Department of Science and Technology to enhance science and engineering education in India over Edusat. It was formally launched on December 8, 2005 when 16 US Universities including Harvard, Yale, Princeton, Georgia Institute of Technology, Michigan, Washington, Illinois-Champaign, Purdue, and Wisconsin-Madison signed an MOU with AmritaVishwa Vidyapeetham.

- **Bharti Airtel Effort:** Has set up in collaboration with IIT Delhi the Bharti-IIT Delhi Centre of Excellence in the Bharti School of Telecommunications Technology and Management at IIT Delhi.

- **Nasscom Initiative:** In collaboration with MHRD, NASSCOM has taken initiatives to help government set up 20 new IIITs on Public-Private Partnership basis in the 11th Plan, has already established Finishing Schools for Graduate engineers to make them more employable, has set up the National Assessment and Competence (NAC) Certification mechanism for converting "trainable" workforce to "employable workforce" in the BPO industry and is starting a higher level accreditation program NAC-Tech for the service sector Industry.

- **INFOSYS Campus Connect Program and Skill Development Initiative:** Campus Connect is Infosys' initiative to help increase India's competitiveness in the knowledge economy. Campus Connect aims at evolving a model through which Infosys and engineering institutions can partner for competitiveness, enhance the pool of highly capable talent for growth requirements in Information Technology (IT) space. It is aimed at creating an effective means of backward integration into the supply chain by going into the college campuses from where the IT industry gets the people for its growth. The goal of Campus Connect is to build a sustainable partnership with engineering education institutions for mutual benefit. There are more than 300 colleges already in the program, they receive faculty training, curricula up gradation advice, student project files and supervision and many other interactive benefits from Infosys. As part of its skill development initiatives, Infosys will train over 6,000 faculty in various engineering colleges/institutes in the next two years (2008-10). The initiative, being taken up under the Indo-US Engineering education program, has already identified 584 engineering colleges (out of 1,500) and 23 senior professors from the US universities who would handle the training.

- **WIPRO Mission 10X:** Wipro Technologies launched its Mission 10X program on 6th September 2007 for creating systematic changes in current teaching-learning practices in engineering education with the hope that by training the faculty along industry standards, the employability level of engineering graduates and diploma holders will improve. Its motto is: "Pursuit of Excellence in Engineering Education through Innovation" It has already released sample resource guides for faculty in Fluid Mechanics and Microprocessors. The sample guides for fluid mechanics and microprocessor would facilitate the delivery of sessions to the learners in a planned and effective manner, people spearheading Mission10X told a press conference here.   The prime objective of Mission10X, is to empower faculty members of engineering colleges to help students imbibe higher understanding of a subject and at the same time develop key behavioral skills that would lead to higher employability. Mission10X program involves a layered set of capability building workshops to help engineering faculty nurture talent pool and it has covered five states till date.

- **TCS-Initiatives:** Tata Consultancy Services (TCS) on 1 March 2008 inked a memorandum of understanding (MoU) with the Indian Institute of Technology (IIT), Guwahati, for setting up a TCS learning centre on the IIT campus and to develop collaboration in research and development based on mutual interests. The TCS learning centre will impart the "Initial Learning Program" for the fresh engineering graduates and the Ignite training program for the science graduates, both on a continuous basis in the proposed facility to be set up. Ignite is a TCS initiative designed to build a strong foundation in the software industry through a customized curriculum. It begins with logic and problem solving methods and moves to their applications in the real world. Initial Learning Program (ILP) is for newly recruited engineering graduates and post graduates in computer application. The centre was scheduled to start functioning in the middle of 2008.

- **Joint TCS-WIPRO-INFOSYS Initiative:** To take the industry-academia partnership in education to the next level, the country's IT majors, including Wipro, Infosys BPO and TCS, are now in talks to come together for education initiatives. So far the three company's had their own educational programs to up grade the quality of IT education in engineering colleges and polytechnics, now they want to synergize their efforts through a new initiative. This initiative is likely to be launched next year and in the first year of its operation it will benefit four lakh students across the 10 universities and the IT giants will adopt their states. Each of the company will adopt a state where they can tie up with the colleges to train their faculty and students.

13.2.3 Institutional Initiatives

- **IIT Bombay Initiatives:** IIT Bombay has launched a major live classroom distance education program IIT-Bombay broadcasts its lectures live through Edusat, the satellite which caters exclusively to the educational sector. Students of any engineering institute not only have real-time access to IIT-B tutoring, but can also interact with resident faculty at Powai. IIT starts transmitting lectures from 8.30 am to 8 pm. Some 100 engineering colleges had already purchased ISRO receivers to access the live IIT-B lectures. To begin with, IIT-B is broadcasting lectures in 13 courses, including software engineering, information systems, computation fluid dynamics, embedded systems, instrumentation and process control and fibre optics communication. Persistent Systems has a tie-up with IIT Bombay to offer a Post-graduate Distance Engineering Education Programs (DEEP) in IT and Management. It offers a distance education program for its employees and other external people who are interested in pursuing these courses. The aim of the DEEP is to provide high quality education to a large number of participants across the country, generating expert IT professionals to meet the requirement of the Indian industry, thereby simulating a classroom environment for students. IIT Bombay has a special MOU with Victoria Jubilee Technological Institute (VJTI) for sharing IIT B resources for post-graduate work of VJTI faculty and students, utilizing CEP and Distance education engineering programs, linking VJTI electronically to receive live lectures from IITB, and improving entrepreneurship training.

- **Andhra Pradesh Institute for Electronic Governance:** The Institute for Electronic Governance (IEG) is a non-profit organization created by Government of Andhra Pradesh, to offer solutions in the discourse of human resources by bringing synergy among the institutions of the government, industry and academia. The main objective is to offer quality human resource and services to IT and ITES sectors. To forge a seamless integration among government, academia and industry, to attain prosperous eco-conscious knowledge society of skilled and empowered students. The Mission of IEG include promotion of synergy among the Institutions of Academia, Industry and Government, imparting employable skills and Industry grade skills to the students and graduated unemployed, developing Skilled human resource through Group Learning following accelerated learning techniques, promoting Curriculum Development to meet the needs of the IT Industry, and Government and promotion of a culture of Research and Development and facilitation of close involvement with the design and implementation of technology solutions for various real life problems. The Programs offered are the Jawahar Knowledge Center (JKC) involved in upgrading the skills of students, the JKC-Infosys campus connect program, and the STP- special training program for SC, ST, BC, and Minority students in collaboration of the state Government, Infosys, JKC and JNTU.

- **NPTEL Initiative:** The main objective of NPTEL program is to enhance the quality of engineering education in the country by developing curriculum based video and web courses. This is being carried out by seven IITs and IISc Bangalore as a collaborative project. In the first phase of the project, supplementary content for 129 web courses in engineering/science and humanities have been developed. Each course contains materials that can be covered in depth in 40 or more lecture hours. In addition, 110 courses have been developed in video format, with each course comprising of approximately 40 or more one-hour lectures. In the next phase other premier institutions are also likely to participate in content creation. Five major engineering disciplines have been covered in this project so far (NPTEL Phase I) at the undergraduate (B.E./B.Tech) level: Civil Engineering, Computer Science and Engineering, Electrical Engineering, Electronics and Communication Engineering and Mechanical Engineering. In addition, a number of core curriculum courses common to all engineering programmers such as mathematics, physics, chemistry, management, electronics, language etc. have also been included. So far 131 Web courses and 113 Video courses have been developed and are freely available for downloading.

- **Georgia Institute of Technology-Hyderabad Campus:** Georgia Institute has signed an MOU with Andhra Pradesh Government on 5th June 2007 for setting up its International campus at Hyderabad and an extension center at Vishakapatnam to be operational by June 2009 to offer its engineering programs in India. It plans to produce some 10-20% of all PhDs in India in the next few years.

- **Manipal Academy Initiative:** Manipal Academy of Higher Education (MAHE) now a deemed University (Manipal University) has established branch Universities at Sikkim and Dubai and signed MOUs with many Industry for employment of their graduates. They attract large number of foreign students in their medical and engineering programs.

- **Networking Engineering Institutions:** Many states have established extensive University electronic networks linking engineering colleges in the State for exchange of course material, video lecturing and conferencing and sharing faculty resources to upgrade the quality of teaching in the connected colleges. Vishveswaraiah Technological University in Karnataka beams engineering courses through EDUSAT utilizing best faculty for the course. MANA TV channels in Andhra Pradesh uses Edusat to beam course material to engineering colleges, polytechnics and Dr. B. R. Ambedkar Open University centers along with similar material to schools, and other degree colleges. It is the largest network of its kind, specifically catering to education and development, in India. The earth station is installed and commissioned at the Dr. B.R. Ambedkar Open University campus and Ku-band receivers are installed at receiving stations.

13.3 KNOWLEDGE COMMISSION RECOMMENDATIONS

The National Knowledge commission was established by the Prime Minister of India in August 2005. The Commission will advise the Prime Minister on matters relating to institutions of knowledge production, knowledge use and knowledge dissemination. The mandate of the Commission is to sharpen India's "knowledge

edge". The Commission will advise the Prime Minister on how India can promote excellence in the education system to meet the knowledge challenges of the 21st Century; promote knowledge creation in S&T laboratories; improve the management of institutions generating Intellectual Property; improve protection of IPRs and promote knowledge applications in agriculture and industry. It will suggest ways in which the Government's knowledge capabilities can be made more effective, making the Government more transparent and accountable as a service provider to the citizen. It will also explore ways in which knowledge can be made more widely accessible in the country for maximum public benefit. The commission has an initial 3-year life and has submitted annual reports in 2006, and 2007 which have made many significant recommendations. The summary of recommendations made on the report of the working group on Engineering Education was sent to the Prime Minister in a letter by the Commission Chairman dated 06 May 2008 (full Text attached in the Annexure 1). There are 9 major recommendations:

- **Reforming the Regulatory Framework:** To establish an Independent Regulatory Authority for Higher Education (IRAHE) with a Standing Committee on Engineering Education under it "to exercise due diligence at the point it approves entry for an institution to grant degree/diplomas" The Committee will determine the criteria and the process of accreditation and license multiple agencies for the same. A mechanism for ranking of institutions would be evolved. AICTE would focus on important issues of curriculum development, pedagogy, faculty development etc.

- **Improving Governance of Institutions:** Progressive dismantling of the affiliating system, grant of full autonomy where feasible, mandatory disclosure on website by all institutions of their infrastructure, faculty, intake and performance of students, recognition status and placement record.

- **Attracting and Retaining Faculty:** Creation of adjunct positions for professionals from industry and research institutions, incentives for attracting and retaining, relaxing PhD requirement for teaching undergraduates, use of ICT and open educational resources (OCR) from the best universities across the world to overcome faculty shortage.

- **Curriculum Reform:** Greater flexibility, interdisciplinary perspective and choice of electives, development of integrating skills such as problem solving and logical reasoning, process orientation, learning ability and communication skills. Industry participation to discuss real life case studies, revamping of laboratory courses, co-curricular activities.

- **Integrating Science and Engineering Education:** Reduction of perceived gap between science and engineering, starting 4-year undergraduate program in science initially in institutes of excellence. Allowing graduates to pursue doctoral program in science and technology without a Master's degree.

- **Encouraging Research:** Vibrant and well funded PhD programs with opportunities for international exposure to attract students who go abroad for PhD, new academic and research institutions to be so located as to help optimize mutual collaboration, setting up of industrial research parks by existing institutions with active research programs, increased resource allocation to Universities to become a hub of research with synergies between teaching and research.

- **Industry-academia Interaction:** Frequent dialog with Government and industry through seminars and workshops to achieve greater alignment of engineering education with employment opportunities, summer internship to be made an integral part of curriculum, use of apprenticeship training for honing the skills of engineering graduates in multiple skills, apprenticeship act of 1961 to be amended if necessary to allow flexibility in training in multiple skills (not a specific trade), continuing education programs for industry personnel, joint institution-industry research efforts to encourage innovation and competitiveness in the global economy.

- **Improve Access:** Government initiative to establish new institutions in the under-provided states, a framework for public-private partnership to be developed in collaboration with concerned states to establish new quality institutions.

- **Mentoring:** Elite Institutions to adopt a few engineering institutions of their choice, and helping them raise standards, making available educational resources in the public domain for use of all students and conducting distance education courses, especially for students at the post-graduate level and working professionals; existing IITs to mentor new IITs, NITs to mentor selected engineering institutions in their region, creating an atmosphere where institutions of distinction feel a sense of calling for the interest of national good.

The above recommendations made in the Chairman's letter to the Prime Minister are based on the detailed Report of Commission's working group on Engineering Education submitted in March 2008. The report has analyzed the current scene and made recommendations for improvement both on short-term and long-term basis. The details of the recommendations of the working group are given verbatim in Annexure 2.

13.4 XITH FIVE-YEAR PLAN FOR HIGHER AND TECHNICAL EDUCATION

13.4.1 Highlights

The following press-release by the Ministry of Human Resource Development on April 1, 2008 highlights the major points in the XIth five –Year Plan for Higher Education including Engineering Education.

1. The Eleventh Plan, as approved by the NDC, identifies "expansion, inclusion and rapid improvement in quality throughout the higher and technical education system by enhancing public spending, encouraging private initiatives and initiating the long overdue major institutional and policy reforms" as the core of the Plan efforts. An outlay of about Rs. 85,000 Crore has accordingly been projected for higher/ technical education during the Plan period, which marks an over 9 times increase (at current prices) over the X Plan outlay for the sector. Priorities for Higher Education in the XI Plan are expansion of access (i.e. of institutional infrastructure), equity (i.e. to ensure effective participation of disadvantaged groups, and to correct regional imbalances), improvement of quality, and optimal use of Information & Communication Technology to promote achievement of these objectives. The XI Plan aims, inter alia, to increase Gross Enrolment Ratio from 11% in 2006 to 15% by 2011-12 through rapid expansion of higher education system while ensuring quality and inclusion, and also restructure and reorient higher education system to meet the requirements of a knowledge economy in a globalized world.

2. These objectives would be achieved by implementing various major programs/scheme in XI Plan which are as under:
    - Establishment of 8 new IITs, of which the process for establishing 4 IITs in Andhra Pradesh, Bihar, Rajasthan and Himachal Pradesh is already underway.
    - Establishment of 7 new IIMs, of which the first, at Shillong, would start functioning from the next academic session.
    - Establishment of 3 new Indian Institutes of Science Education & Research of which one has already started functioning at Mohali and two more would come up soon in Bhopal and Thiruvananthapuram. This is over and above the 2 IISERs at Pune & Kolkata which started in the last year of the X Plan.
    - Establishment of 16 Central Universities in States which have no Central University at present, (viz., Jammu & Kashmir, Punjab, Haryana, Himachal Pradesh, Uttarakhand, Rajasthan, Madhya Pradesh, Chattiisgarh, Bihar, Jharkhand, Orissa, Tamil Nadu, Kerala, Karnataka, Gujarat and Goa), and another 14 Central Universities which would aim to attain world class standards.
    - Establishment of 20 Indian Institutes of Information Technology – as far as possible in the Public-Private Partnership mode.
    - Establishment of 370 Colleges in districts where access and participation rates are lower than the national average. Most of these districts also have a concentration of SCs, STs, and minorities.

- Establishment of about 600 new Polytechnics in Government and PPP sectors and promotion of at least another 400 Polytechnics in the Private Sector. In particular, a polytechnic would be established in every district not having one at present. This would again benefit particularly those districts with a concentration of SC, ST, and minorities, all based on autonomy and marketability principle.

- Establishment of an Indira National Tribal University with countryside jurisdiction to promote study and research into tribal history, economy, society, culture, etc. and into tribal issue, as well as to promote education of Scheduled Tribes.

- Establishment of Women's Hostels for Universities, Colleges and Polytechnics on a large scale.

- Establishment of 2 Schools of Planning and Architecture in Bhopal (Madhya Pradesh) and Trivandrum (Kerala).

- Setting up of 370 degree colleges in low GER districts.

- Identifying 50 Centers for training and Research in frontier areas.

- Supporting uncovered State Universities particularly in underserved regions and for socially disadvantages groups and colleges for quality up gradation to a minimum level.

- Technical Education Quality Improvement Program (TEQIP) Phase II covering inter alia 200 State Engineering colleges,

- Educational loan interest subsidy, scholarship and fellowship schemes.

3. The XI Plan would also accord priority to promoting research of high quality across the Higher Education system, modernizing curricula and evaluation system, enhancing coverage of the accreditation process, and introducing necessary reforms in apex institutions like the UGC and AICTE so as to make them more responsive to current and future challenges and needs.

4. India has been admitted as a provisional member of the Washington Accord during 2007. With permanent membership, Indian students graduating from programs which are accredited by the National Board of Accreditation of the All India Council for Technical Education will have easier access to education and employment opportunities in member countries of the Accord like USA, UK, Canada, Australia, Japan, Korea and Singapore.

5. Special attention continues to be paid to the educational needs of the North Eastern Region. Each State of the region now has a Central University (with Assam having two). Besides the existing IIT at Guwahati, an IIM will, as mentioned earlier, start functioning this year at Shillong. Also, besides the older NIT at Silchar, an NIT has been established in Tripura and establishment of a third NIT in the region (in Manipur) is under active consideration.

6. In the coming years, the thrust in on the use of Information and Communication Technology (ICT) to strengthen the system of Open and Distance Learning. In order to deliver the benefits of ICT in the learning process, a new scheme namely "National Mission in Education through ICT" would be launched, with the objective of providing connectivity to the learners to the 'Knowledge World' in cyberspace and to make them 'Netizens' in order to enhance their self-learning skills and develop their capabilities for on-line problem solving. The Mission would also work for creation of knowledge modules with right contents to address to the personalized needs of learners. It also aims for certification of competencies of the learners, acquired through formal or non-formal means, as also to develop and maintain the database having profile of human resources. In order to deliver the benefits of ICT enabled learning, the Mission would also focus attention on achieving technological breakthrough by developing a very low cost and low power consuming access device, making available free bandwidth, for education purpose, to every Indian, and to build knowledge network between and within institutions, of higher learning in the country.

7. Subject to *the decision of the Supreme Court of India,* initiatives will be taken by the Department of Higher Education to provide for augmenting the number of seats in Central Universities, and Technical Education institutions under the Ministry of HRD to pave the way implementation of the policy on reservation based on the provisions of Central Educational institutions (Reservation in Admission) Act, 2006. The implementation will be carried out in a phased manner over a period of 3 years."

4.2 Some Other Important Recommendations/Observations in the XIth Plan Document Concerning Engineering Education

The complete Eleventh Plan Document is now available on the Planning Commission Website. Education is covered in Volume 2. Extracts concerning Technical Education are reproduced here in Annexure 3. Some of the important observations/recommendations which are not included in the highlights (Sec.12.4.1) are listed below:

● The intake in technical education institutions needs to grow at 15% per annum to meet the skilled manpower needs of the growing economy.

● For establishing the new Technical Institutions proposed (8 IITs, 7 IIMs, 10 NITs, 3 IISERs, 20 IIITs, and 2 SPAs) the scope for Public Private Partnership (PPP) would be explored and for their location and selection of sites clustering will be a key consideration and States would be incentivized for co-locating institutions in strategic locations.

- 7 selected institutions will be upgraded subject to their signing MoUs on commitments for making reforms in governance structure, admission procedures, etc. and aligning with the character of national institutions.

- A quick feasibility study will be undertaken to decide upon the optimum intake capacity of the Central institutions and support them for additional infrastructure.

- Efforts will be made to establish 50 centers for training and research in frontier areas like Biotechnology, Bio-informatics, Nano-materials and Nano-technologies, Mechatronics, MEMS, High Performance Computing, Engineering, etc. However, these will be funded on the basis of specific proposals and on a competitive basis.

- The State Engineering Colleges which suffer from severe deficiencies in academic infrastructure, equipment, faculty, library facilities, and other physical facilities will be supported since top ranking students in entrance examination of the States opt for these institutions in view of relatively low fee structure and government recognition.

- TEQIP Phase II is expected to be substantially enlarged to cover additional 200 State engineering institutions, diversified, made more flexible and allow for greater involvement of States in design and implementation.

- An integrated S&T Plan would be evolved involving UGC, Department of Science and Technology (DST), CSIR, Indian Council of Agricultural Research(ICAR), Departments of Atomic Energy and Space to provide the resources needed for substantially stepping up support to basic research, setting up a national level mechanism for evolving policies, and providing direction to basic research.

- Evolving an empowered National Science and Technology Commission responsible for all matters relating to S&T (Administrative, Financial, and Scientific) including scientific audit and performance assessment of scientists and scientific institutions through peer review.

- The present output of about 450 doctorates per annum in Engineering and Technology, should increase several folds with the expanded technical education capacity, offering substantial scope for postgraduate and doctoral level programs.

- Government schemes and AICTE will proactively encourage establishment of higher (technical) institutions in deficient States.

- The changing role of the Apex Regulating Bodies (UGC, AICTE, MCI, Bar Council, NCTE etc.) will be reviewed in the light of what these organizations are expected to perform in the context of global changes, with a view to enabling them to reach out, regulate and maintain standards, and meet the challenges of diversification to enhance access and maintain the quality and standards of higher, professional, and technical education.

- A knowledge city in Mohali, Punjab is being planned with a vision to promote innovation and startup companies having a cluster of institutions on a single campus: the Indian Institute of Science Education and Research (IISER), National Agri-food Biotechnology Institute, Nanotechnology Institute, Management School, Technology and IP Management Centre, Business Centre, an Informatics Centre, Centralized Platform, Technology facility, a Good Manufacturing Practices (GMP) compliant Bio process Facility for Food and Nutriceuticals, a Technology Park for start-ups, and a host of other shared facilities. Governance, as a cluster is so designed as to allow dynamic contact and collaboration within the cluster and with all existing local institutes and enterprises. Building cluster in strategic location enables innovation and creates necessary synergies and sharing.

- A Faculty Augmentation and Development in Science and Technology initiative is to be started with the following components: a substantial increase in the intake in Junior Research Fellowship (JRF); enhancing research fellowship for PhD students if they are given additional responsibility to also take up teaching as lecturer and making eligible non-NET PhD scholars also for fellowship; increasing the number of fellowships and the quantum of assistance for MTech students; making the teaching system attract and retain the best talent with better pay/perks and funded research, and performance-based rapid career progression; increasing industry–institution interface including provision for tenure jobs in industry for faculty; setting aside a share of project funds as incentive payments for the searchers/fellows; selecting top class institutions to undertake special programs for best faculty development; infusing knowledge capital in the Centers of Excellence through MoU; opening up institutions for international faculty, visiting programs, and faculty exchange; reviewing recruitment policy of faculty for providing more flexibility in appointments, short-term contracts, assignments, and possibility of outsourcing select faculty that is in short supply; initiating a major expansion of faculty development program.

13.5 CONCLUSIONS

The initiatives being taken by the Government, private sector and Institutions mentioned in the above paragraphs confirm the commitment of the Nation in tackling the problems and challenges of Higher and Technical Education. The XI Plan financial allocation for Higher and Technical Education which is a quantum jump from the allocation in previous plan periods and the several innovative proposals included for inclusion in the Plan indicate a firm determination of the Government in meeting the anticipated shortage of trained power demand of the growing Indian economy and the existing shortage of qualified faculty and at the same time making high quality Engineering education of international standard available to all citizen of the country who can benefit from it . One only hopes that the implementation of the Plan proposals keeps pace with the stated intentions.

ANNEXURE-1: National Knowledge Commission Chairman's Letter to the Prime Minister on Engineering Education

Sam Pitroda, Chairman, 06 May, 2008

Dear Mr. Prime Minister,

Engineering education is among the key enablers of growth for transforming India's economy. The quality of teaching and research in this sphere will play a critical role in the emergence of our country as a global knowledge leader. It will also provide vital inputs for enhancing productivity across sectors. In the past two decades, we have seen an eight-fold increase in the number of institutions imparting engineering education at the undergraduate level. Yet, there are some fundamental issues that need to be addressed. A glaring regional imbalance has emerged in the availability of engineering education. Two-thirds of the engineering institutions are located in four southern states, plus Maharashtra, even though they account for less than one-third of the population. There is much less access for the youth in under-provided states, particularly because only 15 per cent of the total seats are available for those who come from outside the state. It would be worthwhile to study whether there are any cultural or region-specific factors that influence the choice of engineering as a career in some states and not elsewhere. This could help make the spread of such colleges more even nationwide. Several recent studies have flagged the problem of unemployability of engineering graduates, largely because curriculum and syllabi are not quite compatible with industry requirements. Further, the standards of a very large proportion of institutions at the bottom of the pyramid have also been found to be abysmal. Even good institutions are plagued by deficiency of quality students at post-graduate and research level. The problems are complex and deep rooted. The situation calls for a new paradigm in regulation, accreditation, governance and faculty development. As part of its consultative process, NKC constituted a Working Group of experts from academia and industry under the chairmanship of Prof. M.S. Ananth, Director IIT Chennai. The names of the members are listed in the annexe to this letter (deleted). The Group has also considered the inputs provided by the earlier committees on this subject chaired by R.A. Mashelkar (1998), U.R. Rao (2003) and P. Rama Rao (2004). The study conducted by Professors Banerjee and Muley of IIT Bombay (2007) has also been taken into account. Based on inputs from the Working Group and consultations with other stakeholders, NKC proposes the following set of initiatives:

1. REFORMING THE REGULATORY FRAMEWORK

As stated in our earlier recommendations relating to Higher Education, there is a need to establish an Independent Regulatory Authority for Higher Education (IRAHE) to cover all streams. The role of the Standing Committee on Engineering Education under IRAHE would be to exercise due diligence at the point it approves entry for an institution to grant degrees/diplomas. The members of the Committee should comprise eminent educationists, education administrators and management specialists drawn from industry. The Committee would follow transparent and uniform processes, under the overall supervision of IRAHE. The Committee shall also determine the criteria and the processes of accreditation and license multiple agencies for the same. A mechanism for ranking of institutions to enable students to take informed decisions at the time of admissions by stipulating grading norms and nominating independent rating agencies also needs to be established. These initiatives will enable the All India Council for Technical Education (AICTE) to focus on important issues such as curriculum development, pedagogy, faculty development etc.

2. IMPROVING GOVERNANCE OF INSTITUTIONS

In order to encourage greater flexibility and autonomy, there is a need to progressively do away with the system of affiliation of engineering institutions/colleges to universities. Where feasible, they should be given full autonomy. To attain greater transparency and accountability, it should be made compulsory for all engineering institutions to display information about their buildings, labs, faculty, intake of students, performance of students, recognition status and placements, on their websites. As emphasised by us repeatedly, appointments of Heads of Institutions must be made through the process of a Search Committee that has an independent Chair and is at an arm's length from the government. The direct involvement of administrative ministries in the process can result in unfortunate situations.

3. ATTRACTING AND RETAINING FACULTY

The most serious challenge in engineering education is the dearth of well-qualified faculty. Several measures must be undertaken in this direction:

- Institutions should be encouraged to create adjunct positions and invite professionals from industry and research institutions to participate in the teaching process.
- The criterion of holding a PhD for teaching undergraduate students may be relaxed to Master's degrees which are specifically designed with more course work in lieu of thesis. Special efforts should be made at the undergraduate

level to identify and motivate those who have the potential as well as the inclination towards teaching.

- Incentives like better salary, modern infrastructure, better living and working environment, possibility of secondment to industry during vacations should be made available.

- Shortage of faculty could also be overcome by innovative use of Information & Communication Technology and Open Educational Resources (OER) by leveraging the content available from the best universities across the world.

- Several initiatives need to be undertaken to improve training and professional development of the faculty. A two-week teacher induction training course should be made compulsory. A one-day regional workshop on teaching/ learning processes should be arranged as a part of the academic calendar. Course development should be made an integral part of the training programmes conducted by Academic Staff Colleges. Better opportunities maybe created for continuing education using the distance mode.

4. CURRICULUM REFORM

The current curriculum should be modified to provide greater flexibility, interdisciplinary perspective and choice of electives. The focus in the teaching/ learning process should be on integrating skills such as problem solving and logical reasoning, process orientation, learning ability, English communication and programming fundamentals. Industry participation to discuss real life case studies should be encouraged. Laboratory courses must be revamped to develop a healthy attitude towards experimental work. Environment must be created to encourage students to participate in co-curricular activities.

5. INTEGRATING SCIENCES AND ENGINEERING EDUCATION

We have entered a period in history where the distinction between sciences and engineering has all but disappeared. Sciences are at the heart of engineering. To that extent there is no distinction between the two. In order to reduce the perceived gap between sciences and engineering we need to create mechanisms that allow mobility between the two streams. One option could be to start four year undergraduate programmes in sciences, initially in institutes of excellence where facilities for science programmes already exist. This would enable pursuing doctoral programmes in science and technology, without a Master's degree.

6. ENCOURAGING RESEARCH

Several initiatives are necessary to promote research in engineering disciplines:

- Vibrant and well funded PhD programmes with opportunities for international exposure should be rolled out to attract students who currently go abroad for PhDs.

- The new academic and research institutions being established should be so located that they can optimise mutual collaboration. Likewise, existing academic institutions with active research programmes should be supported to set up high-tech industrial research parks in the vicinity of their campuses.
- Universities must become the hub of research once again to capture synergies between teaching and research. This will require changes in resource allocation, reward systems and mindsets.

7. INDUSTRY-ACADEMIA INTERACTION

In order to attain greater alignment of engineering education with employment opportunities, frequent dialogue with industry and government through seminars and workshops is necessary. To enhance employability, summer internships should be made an integral part of the curriculum. Restrictive provisions and the elaborate regulatory framework of the Apprenticeship Act 1961 have inhibited industry from adequately using the apprenticeship scheme for honing the skills of engineering graduates. Clauses need to be incorporated in the Act that will enable training in multiple skills (not just a specific trade) and allow entry to and exit from the scheme at different points in the education and career cycle. Likewise, industry needs to encourage continuing education programmes for their employees by collaborating with appropriate educational institutes. Academia and industry should engage in joint research to encourage innovation and competitiveness in the global economy.

8. IMPROVE ACCESS

While the government would need to establish new institutions in the under-provided states, it must be emphasised that the recent proliferation of engineering institutions in the southern states is largely the result of private initiatives. A framework for public private partnerships should therefore be developed in collaboration with the concerned states to establish new quality institutions.

9. MENTORING

Elite institutions should consider some additional responsibilities such as adopting a few engineering institutions of their choice and helping them raise their standards, creating and making available educational resources in the public domain for use of all students and conducting distance education courses, especially for students at the post-graduate level and working professionals. In particular, the existing IITs could mentor the new ones being established. The latter, in due course, could play a similar role vis-à-vis others. Similarly, National Institutes of Technology and Regional Engineering Colleges could play a mentoring role for selected engineering institutions in their respective regions. Mentoring by its definition is a voluntary activity, but if we can create an atmosphere where institutions of distinction feel a sense of calling in the interest of the larger national good, it would transform our education.

We believe that the changes and reforms proposed in this letter are necessary to bring about a qualitative transformation in engineering education to meet present and future needs. We look forward to being engaged in consultations for their speedy implementation.

Thank you and Warm Personal Regards,

Yours sincerely,
Sam Pitroda

Dr. Manmohan Singh,
Hon'ble Prime Minister of India
Indian Institute of Technology, Delhi

ANNEXURE 2: The Recommendations of the National Knowledge Commission Working Group on Engineering Education

1. ACCESS AND INCLUSIVENESS

(a) Government should set up new engineering institutions in the under provided regions.

(b) Government should consider setting aside funds for helping poor students get through the entrance tests and later pay the fees in institutions of the students' choice. The government funds should go to the poor and needy in the form of grants and soft loans for coaching and for fees after admission rather than negotiating with the institutions for lower fees.

2. FACULTY

Increasing the number of faculty

(a) Relaxing the criterion of holding a PhD degree for undergraduate teaching — Though faculty with PhD degrees are desirable, given the current state of affairs, this criterion must be relaxed to faculty holding Master's degree in institutions that only offer undergraduate education. In order to ensure the requisite breadth of knowledge of subjects in such faculty, an alternative approach to the Master's degree is recommended, through exposing students to more course-work in place of the one-year project. In the long run, the faculty currently holding BTech degree should be encouraged to acquire a Master's degree as early as possible.

(b)   Increasing the PhD manpower – Measures should be taken to increase the PhD manpower. Good quality graduate students must be motivated for doing PhD. Attractive incentives through opportunity for international exposure like attending international conferences or exchange programs must be provided. A similar scheme which enables a faculty member to spend two months per year working at the site of his international collaborator should also be put in place. Vibrant PhD programs to attract good students who currently go abroad for doing their PhDs should also be rolled out. Undergraduate in sciences or engineering should be allowed to pursue a PhD program (with scholarship) directly in either sciences or engineering with considerable flexibility irrespective of their background.

(c)   Adjunct/Additional Faculty – Professionals from industry and research laboratories should be invited to participate in the teaching process. Institutions should be encouraged to create adjunct positions for them. The upcoming research institutions should be co-located with academic institutions and vice-versa to facilitate this participation. Existing institutions with active research programs should be supported by the government to set up high-tech industrial research parks in the vicinity of their campuses. Internationally competitive talent must be attracted by providing incentives such as better working environments and globally competitive opportunities. Tap potential faculty at early stage — Those who have potential and inclination towards teaching should be identified at their undergraduate level and motivated to take teaching as a career. Opportunities must exist for these teachers to experiment with innovative content and pedagogy along with the opportunity to pursue for further studies.

(d)   Other incentives – In the absence of research opportunities in undergraduate degree granting institutions, a teacher needs incentives which enable him/her to grow professionally. This could be achieved through provisions like secondment to industry during vacation time enabling a faculty member to improve his or her practical skills.

(e)   Ensuring Quality – Non-government autonomous, transparent and credible mechanisms consistent with international good practices need to be put in place for accreditation process to ensure quality of higher education.

(f)   Continuing Education in Distance Mode — Opportunities should be available for faculty to enhance their knowledge and teaching skills through coursework of a rigorous nature using open distance mode. Institutes of excellence like IITs should be encouraged to make available such courses in all modes.

(g)   The short-term training programs including the induction programs conducted by Academic Staff Colleges set up by UGC should be modified away from the present content-independent model. There is a need to set up training programs that teach teachers on what and how to include in courses.

(h)   A one-day regional workshop on teaching/learning processes by carefully identified good teachers and by educational technology experts should be arranged at least once a year as part of the academic calendar. The workshop should focus on one or two aspects of teaching as an aid to improving the quality of education.

(i)   Teacher induction training dealing with the basics should be insisted upon for anyone who wants to be an engineering teacher. This could be a two weeks contact course conducted by senior mentoring faculty at the national level. It is advisable for premier institutions to put up such teacher training courses for those who cannot attend such in-person programs. Subsequent periodic refresher courses should be organized.

3. TEACHING LEARNING PROCESSES

(a)   In the entire teaching learning process, it is important that teachers should be provided complete autonomy while ensuring accountability by using proper checks and balances.

(b)   In view of changing industry needs, universities should work out a healthy balance between the wholeness of knowledge and specialized courses. Teacher should concentrate on fundamentals, the framework for incorporating empirical knowledge and on imparting problem solving skills. Academic institutions must consider an inclusive approach to integrate skills in the teaching learning process for the holistic development of the individual. Some of these skills include: i. Problem Solving and Logical Reasoning (Analytical Ability); ii. Process Orientation (Attention to Detail); iii. Learning Ability; iv. English Communication (Written and Verbal i.e. non-voice); and v. Programming fundamentals (the generic domain). As far as possible, these skill sets should be integrated as part of all subjects and not taught as separate courses. Industry participation to discuss real life case studies should be encouraged.

(c)   There is a need for introduction of a credit based, semester system in all engineering/science colleges/institutions with a common core of engineering/science in the first two years and a flexible professional stream in the last two years. Transfer of credits from one major to another major within the Institute and also between institutions should be enabled. The common core could consist of courses in mathematics, physics, chemistry, biology, humanities and social sciences, computing skills, communication skills, workshop practices and laboratory practices.

(d)   Curriculum — The current curriculum should be modified to provide flexibility, interdisciplinarity and choice of electives. A self study course to encourage independent study, project work in as many courses as possible to engage diverse groups of students together and one course on current topics of social or industrial or national relevance. A flavour of engineering should be given in the first semester itself. At least one of the Humanities courses should concentrate on developing oral and written communication skills. Laboratory courses should be revamped to include familiarization with standard equipment and measurement techniques and students should be helped to develop a healthy attitude to experimental work. Although voluntary in nature, students at all levels must be encouraged/ motivated/ persuaded to participate actively in co-curricular activities.

(e)   Pedagogy — Where appropriate, teachers should be encouraged to use audio-visual aids including computers in the classrooms to improve communication between them and the students. At least a few modern class-rooms should be set-up in each department. Videotaped feed-back should be provided to help faculty improve their teaching methods.

(f)   Evaluation of Students — Faculty should have the freedom to design their own evaluation systems and experiment with them. However, the evaluation scheme must be clearly spelt out early in the course. It is desirable that students are evaluated continuously for their learning.

(g)   Feedback on Teaching — Teacher evaluation and course evaluation must be conducted scientifically using different feedback forms for theory and laboratory. These processes should be computerized in order to provide early and comprehensive feedback to the teachers.

(h)   Teacher-student and student-student relationships — Teacher student relationships outside the classrooms should be encouraged by providing different structured avenues in which both participate informally.

4. INTEGRATING SCIENCES AND ENGINEERING EDUCATION

In order to reduce the perceived gap between science and engineering, it is desirable to start four year undergraduate programs in science along the lines of engineering programs. These can take off initially in institutes of excellence where facilities for science programs already exist.

5. INDUSTRY INSTITUTE INTERACTION

(a)   In order to meet the increasing demand, more institutes of excellence need to be established. Public private partnership could be explored for the same purpose. However, all polytechnics should be operated in PPP mode.

(b)   Frequent dialogues among Educational Institutes, Industries and Government through seminars and workshops are necessary to keep all informed about the latest trends and issues.

(c)   Academia and industry should engage in joint research in order to improve innovation and competitiveness in the global economy of today.

(d)   In order to solve the issue of employability of graduate, an exposure to industry in form of Industrial Training should be provided to students during their tenure. Finishing schools should be started as a joint venture to train the students in Industry specific skill sets upon graduation.

(e)   The industries may select final year engineering students from different engineering colleges and can offer 'Industry Specific Training' during the vacation to supplement the curricular learning in the colleges.

(f)   The current Apprenticeship Act should be amended to include option of training in multiple skills with provision of multiple entry and exit.

(g)   The Industries can encourage continuing education programs for their employees by collaborating with suitable education institutes.

6. DISTANCE EDUCATION (DE)

(a)   Universities must take full advantage of developments in ICT since the brick-and mortar model is likely to be too expensive – to meet the explosive increase in demand for higher education.

(b)   DE should be used to drive the faculty development program to a great extent.

(c)   Different modes of learning (both offline and online) should be explored in order to work with the existing limitation of suitable infrastructure. In this regard, create a portal based push mechanism to drive continuous learning and build a bank of innovative projects that can be used by the colleges to drive self learning. Feedback mechanism should be established for continuous improvement of pedagogy.

(d)   Work towards quality access by enabling world class infrastructure even in the remote areas of the country.

7. REGULATION

(a)   An apex independent regulatory authority should be established that can achieve the objectives of regulation without political interference. An autonomous Standing Committee for Engineering Education should be established under proposed Independent Regulatory Authority of Higher Education. Its main role would be to exercise due diligence at the point it approves a license to grant degrees/diplomas. The members of such an authority have to be chosen through an international committee consisting of the best educationists, education administrators, management specialists drawn from industry and change management special.

(b)   Establish a system of "Chartered Educationist Certificate" under which
       Chartered Educationist will perform the role of auditing and institutions
       will be mandated to undertake the entire auditing process. In order to ensure
       success, multiple independent agencies should be established.

(c)   Accreditation of Institutions – The Standing Committee on Engineering
       Education shall determine the criteria and the processes of accreditation in
       consultation with experts from academia and industry. It will then license
       multiple agencies which could undertake accreditation.

(d)   Ranking of Institutions – It is essential to introduce ranking of institutions
       based on overall performance as well as sector specific performance based
       not only on technical but also on inclusivity, gender equality, international
       practice, non academic parameters like in USA. The Standing Committee
       will stipulate grading norms and nominate independent rating agencies to
       assess and categorize Engineering institutions.

8. GOVERNANCE

(a)   Autonomy to institutions — The system of affiliation should be done away
       with as soon as possible. Every college should be given full autonomy not
       only for the management, but all the way down to individual teacher so as
       to encourage experimentation.

(b)   Transparency — It should be made compulsory for all colleges to give all
       information about their buildings, labs, faculty, intake of students, output
       performance of students, placements, Chartered Educationist report etc.
       on the website — so that parents, students, consultants, media etc can see
       the trends and make informed decisions. In order to ensure transparency
       on the part of management, a University-Institute Charter (UIC) should be
       set up to provide: transparency in admission and registration in different
       programs of study, an effective teaching and learning environment, guidance
       and supervision through a variety of study methods, tutorial support, student
       academic representation, extracurricular activities, provide institutional
       mechanism for addressing major concerns of students and parents and
       ensure health, safety and security as far as possible. At the same time, a
       Student Charter should also be introduced which clearly spells out standards
       of service expected by and from students.

(c)   Institutes should consider having two cells — Counseling Cell and Training
       and Placement Cell. The former is more of general nature where students
       especially the ones in first year get help to adjust in a new environment. The
       latter is to help graduating students to prepare themselves for the placement
       as well to handle the stress of a failure.

OTHERS

(a) The Government should help the elite institutions maintain their excellence. Policy framework and procedural simplicity should be such as to enable more and more institutions to become elite. At the same time, in such elite institutions, the emphasis should be on global level of excellence in teaching, research, and consultancy. Competition and collaboration between these institutions will lead to many creative and innovative educational methodologies in India and set trends for the other engineering colleges to follow.

(b) Additionally, the elite institutions should consider some additional responsibilities such as (i) adopting a few engineering institutions of their choice and help them raise their standards as part of their corporate responsibility, (ii) creating educational resources for use of students in other engineering colleges and make them available in the public domain, and (iii) conduct distance education courses especially at the post-graduate level for students as well as working professionals.

ANNEXURE 3: Extracts from Planning Commission Document on the XI Five Year Plan (2007-2012)

Volume 2: TECHNICAL EDUCATION Section:
Technical Education: Goals and Targets in Eleventh Plan

1.3.41 During the Eleventh Plan, intake of technical education institutions needs to grow at an estimated 15% annually, to meet the skilled manpower needs of our growing economy.

SCHEMES FOR EXPANSION AND UP GRADATION

1.3.42 The Eleventh Plan envisages setting up of 8 new IITs, 7 new IIMs, 10 new NITs, 3 IISERs, 20 IIITs, and 2 new SPAs. In establishing these institutions the scope for PPPs will be explored. Seven selected technical institutions will be upgraded subject to their signing MoU on commitments to making reforms in governance structure, admission procedure, etc. and aligning with character of the national institutions. In the location and selection of sites for the new institutions, clustering will be a key consideration and the States will be incentivized for co-locating institutions in strategic locations.

EXPANSION OF INTAKE CAPACITY IN THE EXISTING CENTRAL INSTITUTIONS

1.3.43   The recent recommendations of the OSC to increase the intake capacities of the centrally funded technical institutions in the categories of IITs, NITs, IIITs, NITTTRs, and IIMs provide for an opportunity for major capacity expansion of high level technical and management institutions while providing for social equity.

1.3.44   Considering the urgency in expanding the intake capacity due to the acceleration in demand for technical education, a quick feasibility study will be undertaken to decide upon the optimum intake capacity of the Central institutions and support them for additional infrastructure, etc. In view of the increasing demand particularly for MBAs, Departments/ Institutes of Management and Business Administration in the university system will also be strengthened.

STRENGTHENING STATE TECHNICAL INSTITUTIONS

1.3.45   The State Engineering Colleges suffer from severe deficiencies in academic infrastructure, equipment, faculty, library facilities, and other physical facilities. Top ranking students in entrance examination of the States opt for these institutions in view of relatively low fee structure and government recognition. These are supposed to be model institutions for the private sector institutions to benchmark their standards. If standards and norms are insisted upon for private institutions, the government cannot keep its institutions in unsatisfactory condition.

1.3.46   TEQIP Phase II is expected to be substantially enlarged to cover additional 200 State engineering institutions, diversified, made more flexible and allow for greater involvement of States in design and implementation. There will be one-time assistance for project-based support and funds will be released on performance and the State Government accepting a minimum set of reforms including curriculum revision, internal assessments, faculty up gradation, adoption of seminar-tutorials, and the semester system, etc. Proper appraisal system of the projects and effective Monitoring and Evaluation (ME) system will be established. TEQIP-II projects will be on log frame.

1.3.47   Efforts will be made to establish 50 centers for training and research in frontier areas like Biotechnology, Bio-informatics, Niño-materials and Niño–technologies, Macaronis, MEMS, High Performance Computing, Engineering, etc. However, these will be funded on the basis of specific proposals and on a competitive basis.

SCIENCE AND TECHNOLOGY (S&T): THE CUTTING EDGE

1.3.48   In the current knowledge era, our development depends crucially on the ability to harness S&T to find innovative solutions. Capabilities in S&T, therefore, are reckoned as a benchmark for establishing the status of the development of a nation. India must occupy a frontline position in this listing. The Eleventh Five Year Plan approach to S&T will be guided by this ambition and emphasis will be on:

— Evolving an integrated S&T Plan involving UGC, Department of Science and Technology (DST), CSIR, Indian Council of Agricultural Research (ICAR), Departments of Atomic Energy and Space to provide the resources needed for substantially stepping up support to basic research, setting up a national level mechanism for evolving policies, and providing direction to basic research.

— Enlarging the pool of scientific manpower and strengthening the S&T infrastructure. Focused efforts will be made to identify and nurture bright young students who can take up scientific research as a career.

— Promoting strong linkages with other countries in the area of S&T, including participation in mega international science initiatives.

— Evolving an empowered National Science and Technology Commission responsible for all matters relating to S&T (Administrative, Financial, and Scientific) including scientific audit and performance assessment of scientists and scientific institutions through peer review.

— Supporting the schemes suggested by the Empowered Committee on the Science Education.

FACULTY DEVELOPMENT AND RESEARCH

1.3.49   The world over, it is recognized that R&D efforts are imperative for sustained economic growth and social development. However, in India there has been a low level of R&D efforts, mainly due to the inadequate number of highly trained and knowledgeable R&D personnel—particularly at the level of PhDs—relatively low investment in R&D by the corporate sector, and the lack of synergy among R&D institutions and universities. The present output of about 450 doctorates per annum in Engineering and Technology, should increase several folds with the expanded technical education capacity, offering substantial scope for postgraduate and doctoral level programmers.

NATIONAL SCIENCE AND ENGINEERING RESEARCH BOARD (NSERB)

1.3.50   Upgradation of science education and research infrastructure in the universities is a major challenge. The DST would adopt a two-pronged strategy to achieve this objective: (i) expansion and strengthening of S&T base in the universities through appropriate HRD measures and building up of research capabilities of the academic sector and (ii) funding for undertaking internationally competitive and front-ranking major research programs. For this purpose, the existing Science and Engineering Research Council mechanism of the DST would be restructured into NSERB and a special program for rejuvenation of research in universities would be initiated. The proposed Board will address these issues and follow global best practices.

REDUCING WIDE REGIONAL DISPARITIES

1.3.51   Southern States have successfully attracted capital and students from all over the country. Government schemes and AICTE will proactively encourage establishment of higher (technical) institutions in deficient States.

OVERSIGHT COMMITTEE (OSC)

1.3.52   In pursuance of the 93rd amendment to the Constitution of India aiming to provide statutory reservations to SCs, STs, and OBCs in Central Educational Institutions, the Central Educational Institutions (Reservation in Admission) Act has been enacted and has been notified in January 2007. The OSC (Moily Committee), constituted in May 2006 recommended an investment of Rs 17270 crore over a period of five years for the Central Educational Institutions to increase their intake capacity by 54% so as to provide 27% reservation to OBCs without affecting the number of general seats. Of this, Rs 7035 crore will be non-recurring expenditure, the bulk of which will be spread over year 1, 2, and 3, whereas, the recurring expenditure will be Rs 10235 crore spread over five years, increasing progressively subject to the final order of the Supreme Court. An Inter-Ministerial Monitoring Committee will be constituted in the Planning Commission to oversee and review the progress. (See Annexure 1.3.2.)

FEES IN HIGHER EDUCATION, SCHOLARSHIPS, FELLOWSHIPS, AND LOAN SCHEMES

1.3.53   The national commitment 'to ensure that nobody would be deprived of higher education opportunities due to lack of financial resources' necessitates a serious look at the issues of fees, scholarships, and loan schemes.

1.3.54   At present, fees vary across universities, but generally these have been kept very low, in many cases not even covering 5% of the operating cost. The Centre and State Governments must either be able to subsidize university education massively or try to mobilize a reasonable amount from those who can afford it by way of fees that cover a reasonable part of the running cost. Since most university students come from the top 10% of the population by income levels, they would be able to pay fees amounting up to 20% of the operating cost of general university education. The fees for professional courses could be much higher. The fee levels should, therefore, be increased gradually in existing institutions but the new norms could be implemented in new institutions from the start. It may be noted that the new institutions will take time to start.

1.3.55   It must be recognized that there will be some students who cannot afford to pay the increased fees and they should receive scholarships. From a fiscal perspective, the government has to bear the cost either by undercharging fees or providing scholarships. The latter method is most preferable because not all students need scholarships and those that do should be able to avail of the scholarship at any recognized university, thus providing an incentive for universities to compete and attract students rather than have all their costs covered. With a portable scholarship system, the demand for admission in the better universities will signal their preferred standing.

1.3.56   The operating cost of providing technical and medical education is much higher then general education and fees in these institutions will have to be higher. However, these courses also provide opportunities for much higher earnings for most graduates. The additional cost to the student of taking these courses, beyond the basic level of fees referred to above, can be met through student loan **programmes**. Banks are currently providing student loans but there are operational problems. Students at premier institutions such as the IITs or IIMs find no difficulty in getting bank loans, but in other institutions, loans are often linked to providing collateral.

INCREASING AFFORDABILITY THROUGH SCHOLARSHIPS, FELLOWSHIPS, AND LOANS

—   Scholarships to colleges/universities students.

—   Effective fellowship program and substantial increase in coverage of PhD research students under Junior Research Fellowship (JRF).

—   Encourage NET qualified and PhD students to take to research as a career and for creation of intellectual property.

—   Establish interlink of research faculty with research students in universities by offering research fellowships.

—   A framework for facilitating student loans for professional programs including a Higher Education Loan Guarantee Authority for covering bank loans to students of accredited universities.

1.3.57   It is necessary to move to a position where banks will lend freely to students who have achieved admission to certified institutions against a loan guarantee given by a National Student Loan Guarantee Corporation.

REFORMS IN APEX REGULATORY INSTITUTIONS OF HIGHER EDUCATION

1.3.58   The government has created an elaborate institutional arrangement by establishing the UGC as an umbrella organization for coordination and maintenance of standards of higher education, as also other professional statutory councils for regulating professional and technical education and determining their quality and standards. These include AICTE, Medical Council of India (MCI), Bar Council of India, NCTE, etc. These institutions have played an important role in laying down a strong foundation of higher, professional, and technical education and expanding its base throughout the country. However, consequent upon the major structural changes that have taken place during the last 25 years or so in the domestic education system and its growing linkages and involvement with the international education providers, the context of higher, professional, and technical education has undergone a paradigm shift.

1.3.59   It is, therefore, imperative to review the changing role that these organizations are expected to perform in the context of global changes, with a view to enabling them to reach out, regulate and maintain standards, and meet the challenges of diversification to enhance access and maintain the quality and standards of higher, professional, and technical education This would help create and expand the relevant knowledge base from the point of view of the expanding individual entitlements and increasing the capacity of the economy to take full advantage of the domestic and global opportunities.

1.3.60   A high-level committee will be set up to suggest a specific reforms agenda in this context. Similar exercises will have to be carried out with respect to State level institutional arrangements.

BOX 1.3.3 MOHALI KNOWLEDGE CITY — ADVANTAGES OF CLUSTERING

—   It is planned to build a knowledge city in Mohali, Punjab with a vision to promote innovation and startup companies. The cluster includes, on a single campus, the Indian Institute of Science Education and Research (IISER), National Agri-food Biotechnology Institute, Nanotechnology Institute, Management School, Technology and IP Management Centre, Business Centre, an Informatics Centre, Centralized Platform, Technology facility, a Good Manufacturing Practices (GMP) compliant Bio process

Facility for Food and Nutriceuticals, a Technology Park for start-ups, and a host of other shared facilities. Governance, of a cluster is so designed as to allow dynamic contact and collaboration within the cluster and with all existing local institutes and enterprises.

— Building cluster in strategic location enables innovation. Characteristically, in a cluster, research, technology management, investment and business skills, technology incubators and parks for startups are co-located, functionally linked, based on a common vision. The vision of such a cluster is to create necessary synergies and sharing of resources, ideas, and facilities.

BOX 1.3.4 FACULTY AUGMENTATION AND DEVELOPMENT IN SCIENCE AND TECHNOLOGY

— Substantial increase in the intake in Junior Research Fellowship (JRF);

— Enhance research fellowship for PhD students if they are given additional responsibility to also take up teaching as lecturer and make eligible non-NET PhD scholars also for fellowship;

— Increase the number of fellowships and the quantum of assistance for MTech students;

— Make the teaching system attract and retain the best talent with better pay/perks and funded research. Performance-based rapid career progression;

— Increase industry–institution interface including provision for tenure jobs in industry for faculty;

— Set aside a share of project funds as incentive payments for the researchers/fellows;

— Selected top class institutions to undertake special programs for best faculty development;

— Infusion of knowledge capital in the Centers of Excellence through MoU;

— Institutions to open up for international faculty, visiting programs, and faculty exchange;

— Recruitment policy of faculty reviewed for providing more flexibility in appointments, short-term contracts, assignments, and possibility of outsourcing select faculty that is in short supply;

— A major expansion of faculty development program.

REFERENCES

1. National Policy on Education (1986/1992/2000) available at **www.aicte.ernet.in/National_Policy.doc**

2. Establishment of NAAC, NBA (1994) available at www. naacindia.org; and **www.nba-aicte.ernet.in**

3. Technology Vision of India 2020 (1996) available at **www.tifac.org.in/index1.htm**

4. Information Technology Action Plan (1998) available at **www.it-taskforce.nic.in/ prem.htm**

5. Ambani Report on " A Policy Framework for Reforms in Education, April 2000 Available at indiaimage.nic.in/pmcouncils/reports/education

6. New Science and Technology Policy (2003) available at www.dst.gov.in/stsysindia/ stp2003.htm

7. Setting up and Educational Satellite (2003) available at www.isro.org/decu/projects/ medusat.htm

8. Transforming India into a Knowledge Superpower (2003), Strategy for Action: Dr. K. Venkatasubramanian; Vikas Publishing House Pvt. Ltd., 576, Masjid Road, Jangpura, New Delhi-110014.

9. Draft National Biotechnology Plan (2004/05).

10. Setting up of the National Knowledge Commission (2005) available at www.knowledgecommission.gov.in

11. Moily Committee Report (2006) available at oversightcommittee.gov.in/ocrep.pdf -

12. Establishment of Indian Institute of Science, Education and Research(IISER) at Pune and Kolkata (2006-07) available at www.qbtpl.net/bs/FINAL_PROPOSAL_IISER.pdf

13. Establishment of Indo–US Collaboration in Engineering Education (IUCEE 2007): available at www.iucee.org/news

14. Amrita University—US Collaboration (December 2005) available at www.amrita.edu/ indo-us/media.html

15. Bharti Airtel Effort:Bharti Scool of Telecommunications technology and Management, available at www.iitd.ac.in/bsttm

16. Nasscom Education Initiatives available at www.nasscom.in/upload/5216/ July%205%202007%20%20Education%20Initiatives-Final.doc

17. INFOSYS Campus Connect Program available at www. campusconnect.infosys.com

18. WIPRO mission 10X available at www.mission10x.com

19. TCS- Education Initiatives; news.education4india.com/3926/tcs-learning-centre-on-iit-guwahati-campus/

20. Joint TCS-WIPRO-INFOSYS Initiative: available at www. in.rediff.com/money/2007/ oct/10oped.htm

21. IIT Bombay Initiatives available at www.iitb.ac.in/~cep/about/index.html

22. Andhra Pradesh Institute for Electronic Governance available at **www.ieg.gov.in**

23. NPTEL Initiative available at
www.nptel.iitm.ac.in/pdf/NPTEL%20Document.pdf

24. Georgia Institute of Technology-Hyderabad Campus details available at article.wn.com/view/2007/05/26/Georgia_Tech_campus_at_Hyderabad/

25. Manipal Academy Initiative available at **www. manipal.edu.**

14

Technical Education — Policies, Acts, Regulations and Verdicts

"If we can live with the tyranny of over-regulation of our education system, why are we afraid of the tyranny of autonomy of educational institutions?"

— An unknown educationist

14.1 INTRODUCTION

The education system in India presently is one of the most over-regulated systems. Higher education in Engineering and Technology has comparatively a bigger stake with reference to the regulatory frame-work, various Acts, Regulations brought into force through Gazette notifications and numerous court judgments. The profusion of regulatory controls often leads to legal recourse to redress real or perceived grievances. Many of the provisions contained in legislations and Acts bear upon the judicial verdicts, which in turn bear upon the education process. In some court cases the effected parties take recourse to appeal to higher courts or to review petitions.

The policies on education, the actions and interventions of the regulatory authorities under the provisions of the Acts and the gazetted Regulations and the judicial verdicts affect the dispensation of the education in the country and its quality and standard at all its levels. In this Chapter, some policy issues, various provisions of Acts and Regulations and some judicial verdicts have been discussed limiting to higher engineering and technical education to emphasize that a more comprehensive and in-depth analysis is required to re-visit the National Education Policy and to realize the necessity to evolve an altogether new policy, which addresses itself to the contemporary national and international scenario, emerging WTO-GATS regime, mobility of students, scholars and teachers, education delivery systems and education technology tools, structure and overall framework of curricula, governance and management of education and other issues for imparting quality education.

14.2 THE FIRST DEPARTMENT OF EDUCATION

In the British regime, the first department of education was created in 1910 to look after the education in India. The Central Advisory Board of Education (CABE), the most important advisory body of the Govt. of India was established in 1920 and dissolved in 1923 as a measure of economy. It was revived in 1935 and has been in existence since then. After India achieved independence, a full-fledged Ministry of Education was established. The importance of education is enshrined in the Directive Principles of State Policies which states *"the State shall endeavor to provide within a period of ten years from the commencement of this Constitution, for free and compulsory education of all children until they complete the age of fourteen years"*. The Ministry of Education acquired a new name "Ministry of Human Resource Development" (MHRD) in 1985. The present Department of Education is one of the four constituent departments of MHRD.

> *"State education is a mere contrivance for molding people to be exactly alike one another, ...in proportion as it is efficient and successful, it establishes a despotism over the mind, leading by a natural tendency to one over the body."*
>
> — John Stuart Mill

14.3 DOMAIN OF THE GOVERNMENTS ON EDUCATION

The areas in which the Central Government and the State Governments have domain are described in the Constitution in three Lists; the Central List, the State List and the Concurrent List. School education was earlier on the State List meaning that States had the prerogative to manage their education system. In 1976 education was transferred by a constitutional amendment to the concurrent List with the objective to promote meaningful educational partnerships between the Central and the State Governments. The Central Government now makes the National Policies which States have to follow.

Since education is in the Concurrent List, the educational institutions are obliged to bear with the regulations and controls of both the Central and the State Governments. The State Universities and the Central Institutions may not have much problem with this duality of control apart from the usual controls, but it is not so for the self-financed professional institutions.

14.4 NATIONAL POLICY ON EDUCATION

The first National Policy on Education (NPE) was adopted in 1968 in the light of recommendations of the Education Commission (1964-66). A new National

Policy on Education was adopted in 1986. A Programme of Action (PoA) was prepared thereafter. Several large centrally assisted schemes (e.g., Operation Black Board, Education Technology ET, Vocationalization of the Secondary Education, etc.) were launched and AICTE was established (1987) as a statutory body.

Shortly after the adoption of NPE (1986), it was reviewed by Acharya Ramamurthy Committee which submitted its report in December 1990 that was tabled in both the Houses of Parliament in January 1991. In March 1991 CABE considered the manner in which the Report should be processed and set up a Committee to examine the recommendations of the Ramamurthy Committee. Based on Ramamurthy Committee report, NPE (1986) was revised in 1992.

"If it would be wrong for the government to adopt an official religion, then, for the same reasons, it would be wrong for the government to adopt official education policies. The moral case for freedom of religion stands or falls with that for freedom of education. A society that champions freedom of religion but at the same time countenances state regulation of education has a great deal of explaining to do."

— James R. Otteson

14.5 PRIVATE UNIVERSITIES

A Bill called Private Universities (Establishment and Regulation) Bill 1995 was prepared with the object to provide for the establishment and incorporation of self-financing Universities and for the regulation of their functioning. The Bill allowed registered societies, public trusts and registered companies to set up universities. The Bill designated UGC as the "Nodal Agency" to monitor the functioning of private Universities on behalf of the Central Government and provided that the Act of establishment of the University shall be laid in both the Houses of Parliament. Unfortunately the Bill was put in the cold storage due to opposition of political parties from time to time. There is however an urgent need of a consistent and pragmatic Law for establishment and regulation of private Universities. Several States are now establishing private Universities through legislative action. Unlike a State University Act under which, through an amendment of the Act, a new State University can be established, the private Universities, most of which impart professional education, are established by separate Acts. These separate Acts within a State are substantially the same and variations in the Acts of different States are not significant. Some of them give more autonomy, freedom and flexibility to the sponsors of private Universities than do the others. All of them however provide that the private Universities will follow the norms and standards prescribed by UGC, AICTE and other regulatory authorities.

NPE (1986) is silent on the role of self-financed educational institutions because in 1986 the private education system, as it has evolved now, could not be perceived, nor could the compulsions of GATS and WTO that are unfolding now, be visualized. The intake of students in the self financed engineering/ technical institutions in India is nearly 95% of the national capacity but the country does not have a National Education Policy addressing to this large constituency of students in private institutions of the contemporary India and the emerging World order.

In 2003, an expert Committee appointed by UGC to recommend a "Model Act for Universities of the 21st Century in India", prepared a Concept Paper "Towards Formulation of Model Act for Universities of the 21st Century in India". The Concept Paper had a Questionnaire in its Part II. The Paper was circulated to all the Vice Chancellors asking them to have it widely discussed and requesting them to send their responses on the Questionnaire for which a dead line of December 15 2003 was indicated. Nothing seems to have come out of this UGC exercise. There is no evidence that while establishing private Universities, the States have derived any benefits from the UGC exercise for preparing a model Act although UGC Guide lines have been posted on the website. UGC has also posted a Regulation titled UGC (Establishment of and Maintenance of Standards in Private Universities) Regulation 2003. These Regulations have been a subject of debate and have been questioned widely.

> *"The natural liberty of man is to be free from any superior power on Earth, and not to be under the will or legislative authority of man, but only to have the law of nature for his rule."*
>
> — Samuel Adams

14.6 COURT CASES

The over-regulation of the education system in India is believed to be adversely affecting its healthy growth. The complex system of regulations of Higher Education in Universities and other institutions, particularly those that are self financed, has resulted in conflicts on account of multi agency control, high stakes due to heavy investments made by sponsors of private institutions, far reaching short and long term interests of private providers of education that are some times genuine and fair and at other times vested. These have led to numerous court cases generally involving the Governments, UGC, AICTE and various other regulatory bodies on the one hand and sponsors of private institutions on the other. Many court verdicts have profound influence on the education system and its governance and management. Quite often a judgment of a lower Court gets challenged in higher courts. Many cases have got their final verdicts from the Supreme Court which has also heard review petitions. The judgments are invariably reasoned and therefore are often quite long. Some time they are misunderstood and need clarifications, which results in more litigations. A few examples of some famous cases and judgment thereon amply illustrate how they have influenced the education canvas in the country.

"Laws are made for men of ordinary understanding and should, therefore, be construed by the ordinary rules of common sense. Their meaning is not to be sought for in metaphysical subtleties which may make anything mean everything or nothing at pleasure"

— Thomas Jefferson

14.7 MOHINI JAIN VS. STATE OF KARNATAKA

Issues of Rights: The Supreme Court considered the constitutional validity of capitation fees charged by medical colleges in Karnataka in Mohini Jain case in 1992. Under Article 32 of the constitution of India, Mohini Jain, a resident of Meerut, challenged the notification of Karnataka Government permitting private medical colleges in the State to charge the students from outside Karnataka higher tuition fees. The apex Court extensively discussed the constitutional provisions under Article 21 (right to life), Article 38 (State to secure a social order for the promotion of welfare of the people), Article 39(a) and (f) (certain principles of policy to be followed by the State), Article 41 (right to work, to education and to public assistance in certain cases) and Article 45 (provision for free and compulsory education for children).

Right to Education: The apex court held that "The right to education is concomitant to the fundamental rights under Part III of the Constitution. The directive principles, which are fundamental in the governance of the country, can not be isolated from the fundamental right granted under Part II. The educational institution must function to the best advantage of citizens. Opportunity to acquire education cannot be confined to the richer sections of the society. Capitation fee is nothing but a price for selling education. The concept of "teaching shops" is contrary to the constitutional scheme and is wholly abhorrent to the Indian culture".

Obligations of Educational Institutions: The Colleges in Karnataka argued that they did not receive financial aid from the Central or the State Government and they incurred about Rs. 5 Lakhs per student for a five-year MBBS course and hence they had to charge students according to the capital and expenditure. The Court pointed out that when the State Governments granted recognition to the private educational institution, they create an agency to fulfill their obligation under the Constitution. The State discharges its obligations through state-owned or recognised educational education to enable the citizens to enjoy their "right to education".

The Inference: The Court concluded that right to life is the compendious expression for all those rights which the courts must enforce because they are basic to enjoyment of life with dignity. The right to education flows directly from the right to life. The right to life under Article 21 and the dignity of an individual cannot be assured unless it is accompanied by the right to education. The Court held that charging of capitation fee in consideration of admission to educational institutions was a patent denial of a citizen's right to education under the Constitution. The Court, however, exempted foreigners and NRIs form the purview of the judgment. It implied that the institutions could charge higher fee from these two categories. The Court did not point out the options for funding of those institutions since through them, as observed by the Court, State Governments were fulfilling their obligations.

14.8 UNIKRISHNAN VS. STATE GOVERNMENT OF ANDHRA PRADESH

Three Issues: In the famous Unikrishnan case, the petitioners were imparting medical and engineering education in four States, Andhra Pradesh, Karnataka, Tamilnadu and Maharashtra. They raised three issues,: is education a fundamental right under the Constitution; has a citizen of India right to establish and run an educational institution under Article 19 (1) (g) or any other provision of the Constitution; and do the grant of permission to establish and the grant of affiliation by a university make it obligatory for an educational institution to act fairly in the matter of admission of students. Apart from the Karnataka Educational Institutions (Prohibition of Capitation fee) Act 1984, the apex Court took note of the legal and relevant factual position obtaining in the other three States, Andhra Pradesh, Tamilnadu and Maharashtra, observing that all the matters before the Court arose from those four States and that "Notice in these matters were however directed to all the States in the country" but none of them appeared except those four.

The Relevant Acts: The apex Court examined in detail the relevant Acts of the four States and their amendments related to prohibition of capitation fee. The Court dealt with the first issue by observing that right to Education is not stated expressly as a fundamental right in Part III (of the Constitution); the Court does not follow the rule that unless a right is expressly stated as a fundamental right, it can not be treated as one. The Court cited a number of judgments to establish this ruling, e.g., freedom of press, which is not expressly mentioned in Part III, yet it has been read into and inferred from the freedom of speech and expression, and a number of jurisprudence, viz., right to legal aid and speedy trial, right to means of livelihood, right to dignity and privacy, right to health, right to pollution-free environment and so on.

The Scheme: The Supreme Court laid down a Scheme to regulate admission and levy of fee in private unaided educational institutions, particularly those imparting professional education. As an interim measure, the Scheme prescribed an ad-hoc procedure for determination of fee and certain allied matters. As a long term measure, the Scheme required that AICTE, UGC, the Medical Council of India, and the Dental Council would frame suitable Regulations to replace the ad-hoc provisions of the Scheme.

Follow-up Action: Pursuant to the above decision, fees were fixed in various States in private Engineering Colleges at rates that were found suitable by the Courts. Regulations were framed soon thereafter by AICTE broadly incorporating the features of the Scheme. Similarly, UGC also prepared draft regulations on the subject in case of Medical and Dental Colleges. However, rates of fee were laid down by the Court from time to time. The Supreme Court considered the status of implementation of the Scheme, as modified by its various orders in IA No.66 in CWP No.317 of 1993. The orders passed on August 9, 1996 required the Central Government, including MHRD to convene a meeting of all the concerned authorities for evolving a proper fee structure for Medical, Engineering and Dental Colleges throughout the country. Accordingly, two meetings of all concerned, viz., the Ministries of Health & Family Welfare and Law and the Medical Council of India, the Dental Council of India, AICTE and UGC, and representatives of Technical, Higher and Medical Education Departments were held in August and October 1996 in which all the relevant issues were considered. Pursuant to these meetings, the Ministry of Health & Family Welfare took a view that the proposed fee structure for Medical and Dental Colleges could be evolved separately, and that it was not practicable to have a common fee structure for all professional unaided institutions generally.

The Final View: To take a final view in the matter, a meeting of State Higher and Technical Education Ministers was convened by the Union Minister of Human Resource Development in December 1996 wherein it was decided that self-contained policy directions could be issued applicable to the sectors of education falling within the purview of the Ministry of Human Resource Development, namely Higher and Technical Education. In the meeting it was stressed that while AICTE had notified its Regulations and UGC was in the process of doing so, uniform policy directions were still needed to be laid down by the Government to cover all relevant aspects in determining the rates of fee. It was felt that further changes and details could be incorporated in the Regulations in the light of these policy declarations. President of India approved the policy on Fee determination in Private Unaided Institutions in the Higher & Technical Education Sectors formulated on the basis of the principles arrived at in the meetings of the officials of the State Governments of Andhra Pradesh, Maharashtra, Karnataka and Tamil Nadu and later endorsed in the meetings of the Ministers of State Governments for Higher and Technical Education. An outline of the policy is given below:

General Principles: The policy guide-lines aimed at a fee structure in the self-financed institutions, which is fair to all concerned, namely the students and their guardians, the management, faculty members and employees. The bed-rock of the policy was avoidance of commercialization and profit-making while simultaneously ensuring maintenance of standards and upkeep of facilities and assets. For these prime considerations, the broad general principles included prevention of profit making and ensuring, as far as possible, the principle of no-profit no-loss underlined in the Scheme and without diluting the fundamental concern of avoiding commercialization, make allowance in the fee to provide for replacement and up-gradation of facilities under a transparent and intelligible procedure for fee determination. The Scheme brought all private unaided institutions within the purview of the Scheme, not merely confining it to 'Colleges' but Involving the State Governments concerned in the process of fee determination.

Applicability: The policy guidelines apply to self-financing institutions imparting technical education as defined in the AICTE Act including institutions imparting post-graduate education in Management whether by awarding degrees or otherwise, colleges affiliated to the Universities and Deemed to be Universities if such institutions operate on self-financing basis without receiving maintenance grants from the Central or the State Governments.

Admissions: The admission procedure incorporated in the Scheme enables the Commission to make suitable provisions relating to admissions in accordance with the scheme.

Determination of Fee: Fee in institutions imparting technical education are to be determined by State Level Committees comprising as prescribed in the Scheme with the Vice-Chancellor of one of the Universities in the State to be nominated by the State Government concerned as the Chairperson. Fee chargeable by Deemed to be Universities are to be determined by a Standing Committee of the University Grants Commission constituted as prescribed in the Scheme with a Member of the Commission as the Chairperson. Where a Deemed to be University conducts courses in Technical/ Medical Education, a nominee of AICTE/ Medical Council of India/ Dental Council of India as the cases may be, would be a member. The Scheme prescribes the procedure to be adopted by the Fee Committee, how the fee is to be determined, guidelines for maintenance of fee account and other procedural matters connected with levy of fee.

14.9 T M A PAI FOUNDATION VS. STATE OF KARNATAKA (AIR 1994 SC 13)

The Unikrishnan Judgment: In the T M A Pai case the plea of the appellants was that the judgment in the Unikrishnan case adversely affected the autonomy of minority institutions and needed a review. In its judgment, the Supreme Court recognized the role of private initiatives, observing that *"the State with its limited* *resources and slow moving machinery, is unable to fully develop the genius of Indian people"*, held that *"the decision in Unikrishnan's case, insofar as it framed the scheme relating to the grant of admission and the fixing of the fee, was not correct, and to that extent, the said decision and consequent directions given to UGC, AICTE, Medical Council of Indian Central and State Governments, etc., are overruled"* .

Inadmissibility of Capitation Fee: The Court, while forbidding the charging of capitation fee and profiteering by the institutions, opined that *"the decision on the fee to be charged must necessarily be left to the private educational institution that does not seek or is not dependent upon any funds from the Government but the Government can provide Regulations that will ensure excellence in education."* The Court opined that it was not considered profiteering for the private unaided Colleges to have reasonable surplus to meet the cost of expansion and augmentation of facilities. The Court also made some guidelines for admission of students.

"Good judgment comes from experience. Experience comes from bad judgment."

— Bob Packwood

14.10 BHARATHIDASAN CASE: (SUPREME COURT JUDGMENT, 24-902001; CIVIL APPEAL NO. 2056 OF 1999)

In the Bharathidasan case, the Supreme Court looked at the provisions of the AICTE Act in detail and in-depth particularly with reference to the required approvals of AICTE for an institution to start technical education courses or to add new courses. This has been brought out here in view of the importance of the Supreme Court analysis.

The Case: The question of law that arose for consideration in the appeal, Bharthidasan University and Ors vs. AICTE and Ors was whether the appellant University created under the Bharathidasan University Act, 1981, should seek prior approval of AICTE to start a department for imparting a course or programme in technical education or a technical institution as an adjunct to the University itself to conduct technical courses of its choice and selection.

The Petition of AICTE: When the appellant-University commenced several courses in technology, AICTE filed a Writ Petition No.14558 of 1998 before the Madras High Court seeking a writ of mandamus to forebear the University authorities from running/ conducting those technical courses because the University did not apply for and secure prior approval of AICTE for those courses before their commencement as envisaged under the AICTE Act, 1987 and the statutory regulations made thereunder by AICTE, particularly Regulation No.4, which obligated even a University to obtain such prior approval.

The Stand of the Appellants: The appellant-University took a stand that it did not fall under the class of Technical Institution as defined under Section 2(h) of the AICTE Act and consequently, the regulations made for seeking prior approval of AICTE by the Universities to commence a course or programme in technical education or a new department for the purpose, were in excess of the regulation-making powers of AICTE and consequently, were null and void and could not be enforced against the appellant-University to the extent it obligates even Universities to seek and secure such prior approval from AICTE.

The Verdict of the High Court: The learned Single Judge of the Madras High Court accepted the stand of AICTE by applying and following the ratio of the decision of a Full Bench of the Andhra Pradesh High Court reported in M. Sambasiva Rao alias Sambaiah & Ors. vs. Osmania University, Hyderabad & Ors. [1997(1) Andhra Law Times 629] and ordered the cancellation of the admissions made by the University. When the matter was pursued before a Division Bench of the High Court, the learned Judges in the Division Bench also felt convinced of the ratio laid down by the Full Bench of the Andhra Pradesh High Court and rejected the appeal.

Writ Petition in the Supreme Court: The case was taken to the Supreme Court. Since the approach adopted by the learned Single Judge and the Division Bench of the Madras High Court were on the same lines as the one adopted by the Full Bench of the Andhra Pradesh High Court, which the Madras High Court had also followed, the apex Court, referring to the said decision, considered the correctness or otherwise of the ratio in the said decision. The apex Court anlysed in M. Sambasiva Rao (supra), while adverting to the relevant provisions of the UGC Act, 1956, the Andhra Pradesh State Council for Higher Education Act, the A.P. Universities Act, 1991, the AICTE Act and the AICTE Regulations (Grant of approval for starting new Technical Institutions, introduction of courses or programmes and approval of intake capacity of seats for the courses or programmes) 1994.

The Plea of the Appellant: The counsel for the appellant-University, urged that a University like the appellant, as defined under Section 2 (i), did not fall within the definition of a technical institution contained in Section 2 (h) of the AICTE Act and, therefore, equally stood outside the purview of Section 10 (1) (k) of the said Act and consequently not obliged to seek for and obtain prior approval of AICTE for starting a department or introducing new courses or programmers. The regulations framed by AICTE for the same reason insofar as it obligates even Universities to obtain such prior approval, cannot be held to be binding or enforceable against the appellant by the mere fact that the regulation specifically states so, notwithstanding the provisions contained in the Act stipulating to the contrary and any regulation so made would be void and unenforceable. It was also urged that the decision of the Full Bench of the Andhra Pradesh High Court did not lay down the correct position of law and the decisions of this Court relied upon in the said decision really did not lend any support to the principles ultimately laid down therein and, therefore, the Madras High Court ought to have considered the issues independently and not followed the ratio of the Full Bench in M. Sambasiva Raos case (supra). The strong grievance ventilated on behalf of the appellant was that both the Andhra Pradesh and Madras High Courts failed to properly construe the relevant provisions of the Act, applying the correct principles of interpretation and also giving due consideration and weight to the various stipulations contained in Section 10 which made specific reference wherever the universities also have to adhere to the provisions of the AICTE Act, Rules and Regulations. It was also urged that no Rules or Regulations inconsistent with the provisions of the Act could have been either made under the Act or sought to be enforced, legitimately.

The Rebuttal of the Respondents: The counsel for AICTE, while drawing sustenance from the reasoning of the judgment under challenge as well as the Andhra Pradesh case, urged that having regard to the overall functions and powers of the Council under the Act to ensure proper planning and coordinated development of the technical education system throughout the country, the qualitative improvement of such education and regulation and proper maintenance of norms and standards in the technical education system and matters connected therewith envisaged under Section 10 of the Act particularly Section 10 (1) (k) read with Section 20 (1) (b) of the Act, AICTE has pervasive control over Universities also and consequently, the prior approval of AICTE had to be obtained by even the Universities like any other technical institution for starting any new department or institute or commencing a new course or programme in technical education. The totality of the purpose and scheme, claimed to be underlying the enactment is said to confer such sweeping powers over all functional activities relating to technical education and the Universities could not claim immunity from such obligation cast under the Act and the Regulations made by AICTE.

Observations of the Supreme Court: Duly considering the submissions made by both the sides, the apex Court observed that when the legislative intent has specific mention and expression in the provisions of the Act itself, the same cannot be whittled down or curtailed and rendered nugatory by giving undue importance to the so-called object underlying the Act or the purpose of creation of a body to supervise the implementation of the provisions of the Act, particularly when the AICTE Act does not contain any evidence of an intention to belittle and destroy the authority or autonomy of other statutory bodies, having their own assigned roles to perform. Merely activated by some assumed objects or desirabilities, the Courts cannot adorn the mantle of legislature. It is hard to ignore the legislative intent to give definite meaning to words employed in the Act and adopt an interpretation which would tend to do violence to the express language as well as the plain meaning and patent aim and object underlying the various other provisions of the Act. Even in endeavouring to maintain the object and spirit of the law to achieve the goal fixed by the legislature, the Courts must go by the guidance of the words used and not on certain pre-conceived notions of ideological structure and scheme underlying the law.

Mandate of AICTE under the Act: The apex Court considered the mandate of AICTE under the Act to advise the Central and State Governments for ensuring the coordinated development of technical education in accordance with approved standards. The Court took note of the definition of 'Technical education' in Section 2(f) of the Act and various powers and functions of AICTE as also its duties and obligations under Section 10 of the Act, to take steps to fulfill them, including Section 10(1) (k) that provides for grant of approval for starting new technical institutions and for introduction of new courses or programmes and Section 23, which empowers AICTE to make regulations consistent with the provisions of the Act and the Rules. The apex Court also took note that the Act, for all purposes and throughout maintains the distinct identity and existence of 'technical institutions' and 'universities' and it is in keeping in tune with the said dichotomy that wherever the University or the activities of the University is also to be supervised or regulated and guided by AICTE, specific mention has been made of the University alongside the technical institutions and wherever the University is to be left out and not to be roped in, merely refers to only the technical institution in Sections 10, 11 and 22(2)(b).

Empowerment of AICTE and Limitations: The apex Court observed that whereas AICTE is empowered to inspect or cause to inspect any technical institutions under Section 10(1)(p) without any reservation whatsoever, when it comes to the question of universities, it is confined and limited to ascertaining the financial needs or its standards of teaching, examination and research. Section 10(1) (t) envisages AICTE to only advice the UGC for declaring any institution imparting technical education as a Deemed University and not do any such thing by itself. Likewise, clause (u) of the same provision envisages the setting up of a National Board of Accreditation to periodically conduct evaluation of technical institutions or programmes and does not include 'university'.

AICTE Authority over Universities: The Supreme Court found that all these vitally important aspects show that AICTE created under the Act is not intended to be an Authority either superior to or supervise and control the Universities and thereby superimpose itself upon such Universities merely for the reason that it is imparting teaching in technical education or programmes in any of its Departments or Units. A careful scanning of the provisions of the AICTE Act and the provisions of the UGC Act in juxtaposition, show that the role of AICTE vis-à-vis the Universities is only advisory, recommendatory and a guiding factor and thereby sub-serve the cause of maintaining appropriate standards and qualitative norms and not as an authority empowered to issue and enforce any sanctions by itself, except submitting a Report to the UGC for appropriate action. The conscious and deliberate omission to enact any such provision in the AICTE Act in respect of Universities is not only a positive indicator but should be also one of the determining factors in adjudging the status, role and activities of AICTE vis-à-vis Universities and the activities and functioning of its departments and units. The apex Court observed that all these vitally important facets with so much glaring significance of the scheme underlying the Act and the language of the various provisions seem to have escaped the notice of the learned Judges, their otherwise well-merited attention and consideration in their proper and correct perspective. The ultra-activist view articulated in M. Sambasiva Raos case (supra) on the basis of supposed intention and imagined purpose of AICTE or the Act constituting it, is uncalled for and ought to have been avoided, all the more so when such an interpretation is not only bound to do violence to the language of the various provisions but also inevitably render other statutory authorities like UGC and Universities irrelevant or even as non-entities by making the AICTE a super power with a devastating role undermining the status, authority and autonomous functioning of those institutions in areas and spheres assigned to them under the respective legislations constituting and governing them.

Autonomy of Universities: The apex Court finally observed that while stating that autonomy of Universities should not mean a permission for authoritarian functioning, the High Courts by the construction placed by them have virtually allowed such authoritarianism to the AICTE to such an extent as to belittle the importance and elegant role assigned to the Universities in the Educational system of the country and rendered virtually subordinate to AICTE. In our view, that does not seem to be the object of creating AICTE or passing of the AICTE Act. Such construction as has been placed by the Court in M. Sambasiva Rao's case (supra) which found favour of acceptance of the Court in the present case ought to have been avoided and the same could neither have said to have been intended or was ever in the contemplation of the Parliament nor should have UGC and the Universities been relegated to a role subordinate to AICTE. When it is only Institutions other than Universities which are to seek affiliation, it was not correct to state in the decisions under challenge that a University, which cannot grant affiliation to a technical institution, cannot grant the same to itself. Consequently, the conclusions rendered based on the principles for classifying enactments into 'general law' and 'special law' to keep them within their respective limits or area of operation are not warranted and wholly uncalled for and do not merit our approval or acceptance. AICTE cannot, in our view, make any regulation in exercise of its powers under Section 23 of the Act, notwithstanding sub-section (1), which though no doubt enables such regulations being made generally to carry out the purposes of the Act, when such power is circumscribed by the specific limitation engrafted therein to ensure them to be not inconsistent with the provisions of the

Act and the rules. So far as the question of granting approval, leave alone prior or post, Section 10(1)(k) specifically confines the limits of such power of AICTE only to be exercised vis-à-vis technical institutions, as defined in the Act and not generally. When the language is specific, unambiguous and positive, the same cannot be over-looked to give an expansive meaning under the pretext of a purposive construction to perpetuate an ideological object and aim, which also, having regard to the Statement of Objects and Reasons for the AICTE Act, are not warranted or justified. Therefore, the regulation insofar as it compels the Universities to seek for and obtain prior approval and not to start any new department or course or programme in technical education (Regulation 4) and empower itself to withdraw such approval, in a given case of contravention of the regulations (Regulation 12) are directly opposed to and inconsistent with the provisions of Section 10(1) (k) of the Act and consequently void and unenforceable. The fact that the regulations may have the force of law or when made have to be laid down before the legislature concerned do not confer any more sanctity or immunity as though they are statutory provisions themselves.

The Definition under Section 2(h) of the Act: The apex Court further observed that a reading of Section 10 of AICTE Act clearly indicates where the 'universities' are to be included and where they are to be omitted in the provisions of the Act. The definition of a 'technical institution' under Section 2(h) of the Act clearly specifies that a 'technical institution' cannot include a 'University. The clear intention of the Legislature is not that all institutions whether University or otherwise ought to be treated as 'technical institutions' covered by the Act. If that was the intention, there was no difficulty for the Legislature to have merely provided a definition of 'technical institution' by not excluding 'University' from the definition thereof and thereby avoided the necessity to use alongside both the words `technical institutions' and University in several provisions in the Act.

The Verdict: The definition of 'technical institution excludes from its purview a 'University. When by definition a 'University is excluded from a 'technical institution, to interpret that such a clause or such an expression wherever the expression 'technical institution occurs will include a University will be reading into the Act what is not provided therein. The power to grant approval for starting new technical institutions and for introduction of new courses or programmes in consultation with the agencies concerned is covered by Section 10(k) which would not cover a 'University but only a 'technical institution. If Section 10(k) does not cover a 'University but only a 'technical institution, a regulation cannot be framed in such a manner so as to apply the regulation framed in respect of 'technical institution to apply for Universities when the Act maintains a complete dichotomy between a 'University and a 'technical institution. Thus, we have to focus our attention mainly to the Act in question on the language adopted in that enactment. In that view of the matter, it is, therefore, not even necessary to examine the scope of other enactments or whether the Act prevails over the University Act or effect of competing entries falling under Entries 63 to 65 of List-I vis-à-vis Entry 25 of List-III of the Seventh Schedule to the Constitution. The apex Court allowed the appeal at no costs and consequently set aside the judgment under challenge by dismissing the writ petition filed in the High Court. Having regard to the position of law declared by the apex Court, the decision of the Andhra Pradesh High Court reported in M. Sambasiva Rao's case (supra) was also not considered to lay down the correct position of law.

14.11 DISTANCE EDUCATION AND THE DISTANCE EDUCATION COUNCIL

The Gazette Notification of MHRD: The Dept. of Education, Ministry of Human Resource Development, Govt. of India notified in The Gazette of India (No. 44 dated 02-03-1995, April 8th, 1995) that "on the recommendation of the Board of Assessment for Educational Qualifications, the Government of India has decided that all the qualifications awarded through Distance Education by the Universities established by an Act of Parliament or State Legislature, Institutions Deemed to be Universities under Section 3 of the UGC Act, 1956 and Institutions of National Importance declared under an Act of Parliament stand **automatically recognized for the purpose of employment to posts and services under the Central Government, provided it has been approved by Distance Education Council**, Indira Gandhi National Open University, … and wherever necessary by All India Council for Technical Education".

Announcement of DEC: Through an advertisement Distant Education Council announced that "In view of the aforesaid directive of the Government of India, it is **mandatory** for all the Centres/Institutions/Directorates offering programmes through distance mode to apply to the DEC and obtain prior approval before starting a **Centre/Institution/Directorate or programme.** It is informed that any new degree/diploma/certificate offered through distance mode **not recognized by the DEC shall not be accepted for employment in Government Services** and the institution offering such programmes would face de-recognition. The Government is committed to increase GER in higher education but UGC guidelines preventing deemed universities from conducting distance education programmes do not reconcile with GER objectives of the Government. Appropriate action by UGC against erring universities, both Government funded and private, rather than obstructing deemed universities from conducting distance education programmes is required.

The Court Intervention: Issuing an order in March 2006 in favour of Punjab Technical University (PTU) over its dispute with the Distance Education Council (DEC), the High Court of Himachal Pradesh asked DEC not to interfere with the Distance Education Programme of PTU. In its judgment, the Court said that

"Even though the DEC caters to the subject of distance education programme, neither the Indira Gandhi National Open University Act 1985, nor any statutory notification issued there under provides that distance education programme is its exclusive domain or exclusive monopoly.

14.12 PROFILE OF DEEMED TO BE UNIVERSITIES

Power to Award Degrees: The UGC Act 1956 provides that only Universities can award degrees. Consequently an institution which offers degree level programmes must be affiliate to a University. An affiliation implicitly requires that the institution must use syllabi prescribed or approved by the affiliating University, which would also conduct examinations for the affiliated programmes. The systemic weakness of the arrangement is that the course contents, even after becoming out-dated, perpetuate. Those who teach the courses have no opportunity to update the course contents, nor do they have opportunity to evaluate the performance of their students. This arrangement is incompatible with the spirit and ethos of education. As a remedy, a system of granting academic autonomy was adopted. Institutions that achieved good standards of education and acquired reputation and image could be granted autonomy to give them freedom to design and update their syllabi and conduct their own examinations, but the degree would be awarded by the affiliating University

Genesis of Deemed to be University: Based on the Radhakrishnan Committee report, provisions were made in the UGC Act 1956 for declaring higher institutions of learning with quality and high academic standing as deemed universities. Academic autonomy is thus given to institutions through Section 3 of the UGC Act (1956), which provides for another category of institutions, the 'Deemed-to-be University'. The nomenclature itself indicates that an institution of this category is not a University but since it has power to award degree, it is in that sense, distinct from the institutions that are affiliated to a University and have no academic autonomy. Although Deemed-to-be-University is not a University as defined in Section 2(f), it has all the powers of a University. This system of deemed status is unique to India and does not exist any where else in the World.

Earlier Dispensations: Initially a very few prestigious institutions were designated or converted into Deemed-to-be-University. The institutions retained most of their administrative structure, the head of the institutions continued to be called 'Director'. The institutions constituted Academic Councils in line with Universities and also constituted Governing Body/ Executive Council, etc. Earlier the number of Deemed Universities grew but stringent eligibility requirements to become a Deemed-to-be-University kept their number in check. AICTE norms for technical institutions to become Deemed-to-be-University, required that the institution should have a good academic standing, at least four batches of undergraduate and two of post-graduate students must have passed out and the institute should also have a well established Ph.D. program. It meant that the institution should have at least a standing of eight years. The UGC requirement was 10 years. The turning point came when MHRD decided to allow establishment of Deemed-to-be-University de-novo in 1998-99.

Many Aspirants for Deemed to be University Status De-novo: It came to be generally believed that the initial motivation for de-novo provision was to convert IIIT Allahabad, which was not yet ready in 1999 into Deemed-to-be-University. Very soon a number of applications, mostly from ineligible institutions poured in for conversion into de-novo Deemed-to-be-University. Some of them were allowed to and some others were refused permission. A few institutions of the latter group, which considered that they were better than those who were allowed, went to Court. MHRD/UGC kept the pending applications for conversion on hold and a Review Committee was constituted to look at the de-novo provisions. The Committee submitted a report but with the change in Government at the Centre, another review was taken. Eventually new guidelines were notified in 2005 for establishing a Deemed-to-be- University. In recent years the new guidelines have been made as an instrument for questionable approvals of many deemed universities. This growth seems to have accelerated in the last few years. This has resulted in the unjustifiable growth of deemed universities at the cost of quality and the new guidelines seem to accelerate this growth. In the last five years, three committees were constituted for drafting the new guidelines.

Permission to be Known as University and Section 23: Until recently, the deemed universities did not use the word "university" in their names in view of the prohibition of Section 23 of the UGC Act. MHRD has now permitted Deemed–to-be-Universities to call themselves universities in their names. UGC has clarified: *"The prohibition under Section 23 does not apply to the Deemed-to-be-Universities. Further, all universities, whether Central or State universities or universities under section 3 of the UGC Act should clearly state the provisions under which they have been established and this direction is mandatory."* Section 23 of the UGC Act however states under the title *"Prohibition of the use of the word 'university' in certain cases"* that *"No institution, whether a corporate body or not, other than a university established or incorporated by or under Central Act, a Provincial Act or a State Act, shall be entitled to have the word 'university' associated with its name in any manner whatsoever."*

The Earlier View of the Apex Court: In 1984 the Supreme Court had observed: "It is for the Centre to consider whether an institution covered by Section 3 of the Act would not satisfy the provisions of section 23 of the Act and if in the opinion of the Centre, such an institution is not covered, whether an appropriate amendment to Section 23 should not be made so as to exclude recognized institutions under Section 3 of the Act from the field of prohibition covered by section 23 of the Act."

Profusion of Deemed to be University: Whereas during the five years of UPA Government, legislation both to regulate private universities and open the education sector to foreign investment were continued to be stalled, a very large number of institutions, mostly in the private sector, were granted Deemed-to-be-University status. The deemed status is considered coveted as it allows institutions to be largely free of government interference. The Government has in effect enabled private players to flourish, even as the Private Professional Educational Institutions (Regulation of Admission and Fixation of Fee) Bill, 2005, still remains in the pipeline.

Until the provisions of de-novo Deemed-to-be Universities were made the number of such universities was small, merely 29 in 1991, 35 years after the UGC Act was passed. They included reputed institutions such as IISc, Bangalore, BITS, Pilani and Thapar Institute, Patiala. After the 'de novo' provision was introduced by UGC in 2001, about one third of the many newly created deemed universities were recognized through the "de novo" clause, which benefitted institutions barely a few years old, subject to revision after five years.

Statistics of Growth: In the 1990s, few institutions received the recognition as deemed universities but the trend picked up again in 2002, as the number of deemed universities in India exceeded 50. During the six years of the National Democratic Alliance Government, more than 30 institutions were granted deemed status. Deemed universities more than doubled. At the same time the number of state-run universities saw slower growth (178 in 2002 to 217 in 2006). The growth of Central Universities was very low (18 to 20), until 15 Central Universities were established, 13 of them ab-initio. The recent rapid growth in the number of deemed universities in India indicates a new liberal policy of the UGC, which earlier prescribed strict guidelines for the institutions seeking deemed status. Encouraged by the more than hundred institutions, mostly private, have approached UGC seeking deemed university status. Out of them, nearly one third are seeking the deemed status under the de novo category. With such relaxed guidelines, UGC could be risking commercialization of education. But under heavy demand for higher education in the country and in view of the supporting recommendations of the Planning Commission in favour of financial assistance to deemed universities, UGC is recognising deemed status in growing number. The number of Deemed to be Universities as in June 2008, was127; Tamilnadu topping the list with 29 (nearly half of them in the last three or four years), Maharashtra a close second with 21 and Delhi in double digit, 12. Many Deemed to be Universities have a single discipline programme for award of in-house Ph.D. degrees. Some of the institutions recognised as Deemed to be University with one or two disciplines, include three Foundations (Andhra Pradesh), a Yoga Institute (Bihar), an institute of Foreign Trade, National Museum Institute, a School of Drama, and two Sanskrit Institutes (Delhi), a Brain Research Centre and a

Dairy Research Institute (Haryana), a Yoga Institute (Karnataka), a Society, an Institute for Population Sciences and a Fishery Institute (Maharashtra), and many other single discipline setups; they are in the August company of Homi Bhabha National Institute in BARC and TIFR (Maharashtra), Institute of Space Science and Technology (Kerala), and some others. During the five years of the first regime of UPA nearly 70 Deemed to be Universities were established, many of them de-novo, much like the spate of Chhattisgarh at the turn of the century. Once again the provision of establishing Deemed to be Universities is under review and all the applications of the institutions for this privileged status is on hold

> *"State education is a mere contrivance to molding people to be exactly alike one another,…
> in proportion as it is efficient and successful, it establishes a despotism over the mind,
> leading by a natural tendency to one over the body."*   — John Stuart Mill

14.13 DEEMED TO BE UNIVERSITIES AND THE SUPREME COURT JUDGMENT IN THE CASE OF PRIVATE UNIVERSITIES IN CHHATTISGARH

The Scope of Section 2(f) of the UGC ACT 1956: Clause (f) of Section 2 of the UGC Act 1956 defines University as, *" 'University' means a University established or incorporated by or under Central Act, a Provincial Act or a State Act and includes any such institution as may,* **in consultation with the University concerned***, be recognized by the Commission in accordance with the regulations made in this behalf under this Act".*

The Logical Meaning: The obvious meaning of the words *"the University concerned"*, which UGC is obliged to consult, has to be the University to which the institution that is a candidate for itself to be recognized by UGC as a University, is affiliated. According to UGC Act, such a candidate institution to become a University need not be established or incorporated by or under a Central Act or a State Act (there are no Provincial Acts any more); it only needs to be *recognized* by the Commission in accordance with the regulations made in this behalf under the UGC Act. If the UGC has this latter power under Section 2(f) to recognize an institution as a University, it has never used it. It has however, used very liberally, particularly during the recent years, its power under Section 3 to create Deemed to be Universities.

Deemed to be Universities: Section 3 of the UGC Act provides that *"The Central Government may, on the advice of the Commission, declare, by notification in the Official Gazette, that any institution for higher education, other than a University, shall be deemed to be a University for the purposes of this Act, and on such a declaration being made, all the provisions of this Act shall apply to such institution as if it were a University within the meaning of clause (f) of section 2".*

It may be noted that Deemed to be Universities are also not established or incorporated by or under a Central Act or a State Act. The UGC Guidelines 2000 for considering proposals for declaring an institution as deemed to be University under Section 3 of the UGC Act, in its Para 3 provides that *"to qualify for recognition of status as a university, the institution should have among its primary objectives, postgraduate instruction and training in such branches of learning as it may deem fit, and research for the advancement and dissemination of knowledge."*

De-novo Deemed to be Universities: Para 16 of these UGC Guidelines provides that, *"De-novo institutions in the emerging areas with the promise of excellence, not yet fulfilling the prescribed guidelines of the UGC, may be inspected by a Committee of the UGC for recommending to the Government of India for granting them provisional status of deemed to be university, subject to its confirmation after five years on the basis of performance report of the UGC Review Committee done annually for a five year period. The conditions prescribed under clauses 3, 4(a)(i), 4(e), 8 and 9(a) to (d) shall however, not apply to De-novo Institutions in the emerging areas with the promise of excellence, not yet fulfilling the prescribed guidelines of the UGC."*

A Sweeping Decision, Promising Institutions should not Wait: In support of the aforementioned stipulations of Para 16 of the Guidelines, the then Secretary MHRD in the meeting of the Committee constituted for the purpose of looking into the matter of the so called 'De-novo' Deemed to be University argued, by giving an example of the Indian Institute of Information Technology (IIIT) Allahabad, that why IIIT, for which the Government had ambitious plans, should wait for several years to fulfill the requirements as stipulated by the then AICTE norms, or UGC requirements, before becoming a deemed to be University.

Response to UGC Guidelines: The inevitable happened in response to UGC Guidelines of 2000 making a provision of De-novo deemed to be University. A large number of proposals from Institutions poured in for becoming deemed to be University. This was due to the urge of those Institutions to acquire academic autonomy and freedom from the control of the affiliating University which many Institutions found stifling. The applications were kept on hold pending review of the De-novo Guide-lines. After the review, a rider was introduced to enable "De-novo institutions in the emerging areas with the promise of excellence, not yet fulfilling the prescribed guidelines of the UGC, may be inspected by a Committee of the UGC for recommending to the Government of India for granting them provisional status of deemed to be university, subject to its confirmation after five years on the basis of performance report of the UGC Review Committee done annually for a five year period.

Quality Concerns for Deemed to be Universities: The Report of the Review Committee of AICTE (U.R. Rao Committee) constituted by MHRD which was submitted in September 2003, states in Section 4.1.9 that "The AICTE has laid down a detailed procedure for advising UGC for declaring any institution imparting technical education as Deemed to be University. There are over 45

Deemed Universities in existence as of 2003. However, the AICTE has no hold on these institutions after they obtain autonomous or deemed status, which negates the role of the AICTE of maintaining norms and standards There is neither any independent check on the institutions which have attained deemed or autonomous status, nor has the Review Committee seen any evidence of such institutions having a transparent and rigorous independent audit on the quality of education they are imparting and research they are carrying out. Such institutions have an extremely unsatisfactory state of affairs, which needs to be rectified in the interest of ensuring high quality education. Application of rigorous criteria for recognition of deemed status and mandatory requirement of periodic validation, at least once in five years, of the quality of education imparted by such institutions are needed as elaborated in Section 2 of Chapter-7."

Gazette Notifications for Deemed to be Universities and Chhattisgarh Universities: Most of the Deemed to be Universities have been established in recent years. Under Section 3 of the UGC Act, all the Deemed to be Universities have been established by the Central Government by Gazette notifications from time to time, under the umbrella of the UGC Act, 1956, and all the provisions of the Act apply to each one of them *"as if it were a University within the meaning of clause (f) of section 2"*. Earlier an institution declared as a Deemed to be University did not use the word "University" in its name and continued to be known by its original name. The Head of the institution also did not call himself as Vice Chancellor and continued to use his original designation. But under a new dispensation of MHRD, the Deemed to be University can use "University" in their names. Many Deemed to be Universities, particularly the new ones, have changed their original names and have started using the label "University" in their names. The Chhattisgarh Niji Kshetra Vishwavidyalaya (Sthapana Aur Viniyaman) Adhiniyam which was published in the Gazette on 4-2-2002 to establish self-financed private universities is an umbrella Act. Under the Chhattisgarh Act, 112 private Universities were established in Chhattisgarh State within a short span of time by Gazette notifications in accordance with the provisions of the Act. Several private Universities started functioning and admitted thousands of students in Chhattisgarh and also in their off-campus Centres in the State as well as out side the State. A large number of these private Universities were not viable.

Chhattisgarh Universities Challenged: Establishment of the private Universities in Chhattisgarh was subsequently challenged through the Writ Petition (civil) 19 of 2004 by Prof. Yashpal & Ors vs. State of Chhattisgarh & Ors. The State filed its brief claiming that the Act was passed to facilitate establishment of private Universities with a view to create supplementary resources for assisting the State Government in providing quality higher education.

Sections 5 and 6 of this Act were declared to be ultra vires and were struck down by the Supreme Court in its Judgment of February 11, 2005.

Section 5 of the Chhattisgarh Act made two provisions:

(1)   "The State Government may by notification in the Gazette establish a University by such name with such jurisdiction and location of campus as may be specified therein —."

(2)   "Every notification issued under subsection (1) shall be laid on the table of the Legislative Assembly"

Section 6 of the Act also made two provisions supplementing Section 5:

(1)   Every University established under sub-section (1) of Section 5 shall be a body corporate by the name notified under the said section having perpetual succession and a common seal, and may sue or be sued by the said name.

(2)   The University established under sub-section (1) of Section 5 may, with the prior approval of the State Government, affiliate any college or other institution or set up more than one campus.

As a consequence of the above judgment, the court held that *"all notifications issued by the State Government in the Gazette in the purported exercise of power under Section 5 of the aforesaid Act notifying the Universities (including respondent nos. 3 to 94) are quashed and such Universities shall cease to exist."*

The declaration of Sections 5 and 6 of the Chhattisgarh Act as ultra vires by the Supreme Court means that a State Government has no legal authority to establish a University by a Gazette notification under an umbrella Act of the State Legislature.

Deemed to be Universities in the Same Class: Deemed to be Universities in accordance with Section 3 of the UGC Act for all purposes, are Universities having all the power and privileges of a University which is established by an Act of Parliament or an Act of a State Legislature. The Deemed to be Universities with the power to award degrees, diplomas and other distinctions, are being created by Gazette notifications and not by an Act of Parliament. Many deemed to be Universities recently created through Gazette notifications had Chancellors, Pro-Chancellors and Vice Chancellors like the Universities created by Acts of Parliament and State Legislatures have, even before the new dispensation of MHRD came permitting the Deemed to be University to use "University" in their names.. It is interesting to note that an IIM is neither a University nor a Deemed to be University. IIM's therefore cannot award degrees, they award only Diplomas. On the other hand, IITs have been declared as 'Institutes of National Importance' by an Act of Parliament. They award degrees in that capacity and can be regarded as Deemed to be University. The Regional Engineering Colleges, now National Institutes of Technologies, were made deemed to be University by Gazette notifications. They now have freedom from the requirement of affiliation and they can award degrees. However, presently the majority of Deemed to be Universities are those that were earlier private institutions.

The Parallel between Deemed to be Universities and Chhatisgarh Universities: There are at least two parallels between 112 Chhattisgarh Universities, and the numerous recently established Deemed to be Universities. Both were established in large numbers in a short span of time with little regard to quality, norms and standards in most of them in both cases. The other commonality is in the realm of the Law. Chhattisgarh Universities were established by gazette notifications under an umbrella Act passed by the State Legislature. Deemed to be Universities have been and are being established also by Gazette notifications under the umbrella UGC Act 1956. In the case of Chhattisgarh, the Supreme Court declared the two enabling provisions of the Chhattisgarh Act, Section 5 and Section 6, as ultra vires and struck them down. Maharashtra and Rajasthan also had Acts similar to the expunged Chhattisgarh Act. In view of the Prof. Yashpal's writ petition, Maharashtra kept its Act in abeyance. Rajasthan however, made an amendment to safeguard its Act by amending it and making a provision that the Universities would be allowed to start functioning after the decision of the Government for establishing the University is tabled before the Legislative Assembly and is approved by the Legislature. The amended Act was challenged in the Rajasthan High Court. The High Court upheld the amended Act. This led to an appeal against the High Court judgment in the Supreme Court. Rajasthan Government then followed another route same as that in Uttar Pradesh and Himachal, Uttarakhand, etc. The State Governments now do not have an umbrella Act to establish private Universities. They now pass an individual Act for each University.

Since Education is a concurrent subject, it remains to be explicitly clarified that Deemed to be Universities to which, for the purpose of the UGC Act, all the provisions of the Act under Section 2(f) apply, can be created by the Central Government through a Gazette notification on the basis of an umbrella UGC Act, but the same action taken by a State Government to establish a private University through a Gazette notification on the basis of an umbrella Act passed by the State Legislature, is ultra vires.

14.14 CIVIL APPEAL NO. 5041/ 2005 P A INAMDAR & ORS VS. STATE OF MAHARASHTRA & ORS

The Background: Leave was granted by the Supreme Court to all (48 in number) the Special Leave Petitions (SLPs). The Court observed that a Bench of 11 Judges heard and pronounced the verdict on the case TMA Pai Foundation vs. Karnataka State (2002) 8 SSC 481. It was expected that the authoritative pronouncement of a Bench of such strength on the issue arising before it would draw a final curtain on those controversies. This did not happen. The High Courts were soon flooded with writ petitions calling for settlements of several issues which were not yet resolved or which arose post-Pai Foundation case. Several SLPs against interim orders of High Courts and a few writ petitions directly in the Supreme Court were filed. A Constitution Bench sat to interpret the 11-Judge Bench decision in the Pai Foundation case, which it did vide its judgment (14.8.2003), reported as 'Islamic Academy of Education & Anr. Vs. State of Karnataka & Ors., (2003) 6 SEC 697, Islamic Academy'. The 11-Judge Bench in the Pai Foundation case delivered five opinions. The events following Islamic Academy judgment showed that some of the main questions remained unanswered even after the exercise taken by the Constitution Bench for interpretation of 11-Judge Bench decisions. A few of those questions were considered by the 7-Judge Bench which heard the Inamdar vs. Maharashtra case. Whereas Questions 1, 3(b), 4, 5(a), 5(b), 5(c), 8, 9, and 10 and 11 jointly, were answered, the Questions 2, 3(a), 6(a),6(b), 6(c) and 7, were not, since the Bench decided that they be dealt with by a Regular Bench.

The Questions: The first category of questions that were answered, referred to the meaning and content of minorities (Article 30 of the Constitution); scope of Article 30 for professional education; regulation of admission by the State Government or the affiliating University; minorities' right to establish and administer educational institutions and admissions therein; scope of regulatory control; ratio laid down by the Court; decision of the court in Unnikrishnan case regarding primary education to be a fundamental right; non-minorities right to establish and administer educational institutions (Articles 21, 29(1) read with Articles 14 and 15(1)); meaning of 'education' and 'educational institutions'. The second category of questions that were left out for the Regular Bench, related to the meaning of 'religion' (Article 30(1)); indicia for treating an educational institution as a minority institution; location and jurisdiction of minority institution; eligibility of members of minority on the basis of the location of their residence;

Three Issues and the Orders of Reference: The 7-Judge Bench listed three issues for decision, (i) the fixation of 'quota' of admissions/students in respect of unaided professional institutions; (ii) the holding of examinations for admissions to such colleges, and (iii) the fee structure. The Bench spelled out four questions by two Orders of Reference for decision. The first two were with reference to unaided (minority or non-minority) educational institutions. They related to the extent to which the State can regulate the admissions, policy of reservation and/ or quota in admissions, the admission procedure. The remaining two were for the Islamic Academy. They related to regulation of fee and the admission procedure. The Bench made reference to various Constitutional provisions and several court judgments delivered earlier. The Bench held that the scheme evolved for setting up the two Committees for regulating admissions and determining fee structure by the judgment in Islamic Academy case can not be faulted either on the ground of infringement of Article 19(1) (g) in case of unaided professional educational institutions of both categories and Article 19(1) (g) read with Article 30 in case of unaided professional institutions of minorities.

The Verdict and the Judicial View: Finally, the Bench observed that certain recitals, certain observations and certain findings in Pai Foundation case are contradictory inter se and 'such conflict can only be resolved by a Bench of a coram (quorum) larger than (that of the Bench of) the Pai Foundation.' The Bench pointed out that there were several questions which remained unanswered and there were certain others, which arose post- Pai Foundation and Islamic Academy cases. To the extent the area is left open, the Benches hearing individual cases after the judgment would find the answers. Issues referable to those areas, which are already covered by Pai Foundation and yet open to question, shall have to be answered by *'a Bench of a larger coram than Pai Foundation.'* The Bench left those issues *'to be taken care of by posterity.'* The Court further opined that it is for the Central Government or for the State Governments, in the absence of a Central legislation, to come out with a detailed and well thought out legislation on the subject. Such a legislation is long awaited; States must act towards this direction. Judicial wing of the State is called upon to act when the other two wings, the legislature and executive, do not act. The Bench observed that 'earlier the Union of India and the State Governments act, the better it would be.' The Committees regulating the admission procedure and fee structure shall continue to exist, but only as a temporary measure ... until the Central Government or the State Governments are able to devise a suitable mechanism and appoint competent authority in consonance with the observations made by the Bench and cautioned that any decision taken by such Committees and by the Central or the State Governments, shall be open to judicial review in accordance with the settled parameters for the exercise of such jurisdiction.

14.15 CLOSURE

The Symbiotic Weakness: The natural consequences of too many regulations and too much control are impediments to growth, progress and advancement of what is regulated. Regulations and controls in education system are however, necessary to keep essential order and ensure focus on the national interest and priorities. In the hierarchy, policies are most important on the basis of which plans and programmes are prepared. Acts and Regulations are to be in harmony with the policies and programmes. In this Chapter, some selected Court cases mostly involving professional institutions, the judicial opinions contained therein and the verdicts are compiled to exemplify the symbiotic weaknesses in the education system. Primarily the systemic weaknesses and the education structure for which the pervasive perception is that it must be propped by a plethora of Regulations without which it can not stand, have resulted in numerous Court cases and verdicts. The views, with reference to an important legislation, expressed by the 7-Judge Bench of the Supreme Court in the P A Inamdar's case that "Such a legislation is long awaited; States must act towards this direction. Judicial wing of the State is called upon to act when the other two wings, the legislature and executive, do not act." is just one example of wake-up calls from the apex Court.

The Maze: The matrix of judicial opinions, observations, directions and orders along with the provisions of Acts, Rules and Regulations of the Central and State Governments lead to a formidable maze in which, as the Regulators believe, quality education under the umbrella of regulatory authorities can be provided through more than two thousand technical education institutions to all who desire and aspire for good education. A soul searching is required to ascertain how this can be done. Yashpal Committee, and the Knowledge Commission before him, have suggested scrapping off UGC and AICTE which may be replaced by a new Regulatory Authority. Great foresight is required to replace over-regulation by super regulation that is designed to neutralize the maze.

"The governance of higher education in the 21ˢᵗ century needs to develop a fusion of academic mission and executive capacity, rather than substitute one for the other."

— Institutional Management in Higher Education (OECD)

REFERENCES

1. The University Grants Commission Act, 1956.
2. The All India Council for Technical Education Act, 1987.
3. Supreme Court Cases 666, AIR 1992 SC1858; Ms Mohini Jain vs. State of Karnataka and Ors.
4. Supreme Court, Writ Petition (C) No. 607 of 1992 with others; J P Unnikrishnan and Ors vs. State of Andhra Pradesh and Ors.
5. Supreme Court Cases AIR 1994 SC 13; T M A Pai vs. State of Karnataka.
6. Supreme Court Case AIR 2001 SC 2867; Bharathidasan University vs. All India Council of Technical Education.
7. The Chhattisgarh Niji Kshetra Vishwavidyalaya (Sthapana Aur Viniyaman) Adhiniyam 2001 (No.2 of 2001). (The Chhattisgarh Private University Act 2001)
8. Supreme Court Judgment, October 31, 2002; TMA Pai and Ors vs. the State of Karnataka and Ors. and August 14, 2003.
9. UGC Guidelines for Declaring Deemed to be University.
10. Concept Paper "Towards Formulation of Model Act for Universities of the 21ˢᵗ Century India", circulated by UGC in Oct. 2003
11. Supreme Court Cases 773 (2003) 6; Islamic Academy of Education vs. State of Karnataka
12. Supreme Court, Writ Petition (civil) 19/2004; Prof. Yashpal and Anr vs. State of Chhattisgarh and Ors.

13. Lok Sabha Debates (Part II Proceedings other than Questions and Answers), March 21, 2005.

14. Rajasthan High Court Writ Petition No. 3964/2005; Prakash Chaturvedi and Anr vs. State of Rajasthan and Anr.

15. Various UGC Regulations.

16. Supreme Court, Civil Appeal No. 5041 of 2005 (Arising out of SLP (C) No. 9932 of 2004); P A Inamdar & Ors vs. State of Maharashtra & Ors.

17. The Madhya Pradesh Niji Vishwavidyalay (*Sthapana Avam Sanchalan*) Adhiniyam 2007 (no. 17 of 2007). (The Madhya Pradesh Private University Act).

18. Internet.

15

Consolidated Recommendations

"Extraordinary people visualize not what is possible or probable, but rather what is impossible. And by visualizing the impossible, they begin to see it possible."

— Louis Braille

Based on the detailed analysis carried out in the main chapters of the study, the following major recommendations emerge.

15.1 GENERAL RECOMMENDATIONS

— IITs have been role models of quality institutions in the country since Independence. Time has come for policy makers to look beyond IITs as was recommended by the Anandakrishnan Committee especially since new IITs requiring huge investments are being established. These new institutions need not be carbon copies of the IITs which were established fifty years ago.

— Engineering graduates today require not only adequate technological ability and problem solving skills, but must also be endowed with soft skills like cooperative working, communication and presentation skills, business ethics, and inter-personal relationships, and possess a deep commitment to safety, reliability, quality and sustainability of all engineering activities in which they take part. Engineering Institutions have now a new responsibility to provide opportunities to every student to acquire these abilities in addition to their technical knowledge.

— With an eye on permanent membership of the **Washington Accord**, and the pressure of Internationalization of engineering education, India must line up its accreditation and quality assurance policies in tune with those now prevalent in the rest of the developed world so as to accelerate mobility of Indian engineers in all parts of the globe and help export Indian engineering services.

— To achieve global excellence Indian engineering education must incorporate global competence as a key qualification of engineering graduates, give priority to transnational mobility of engineering students, researchers and professionals, develop strong links with professional practice and engage in engineering research in a global context.

— The Government will have to get actively involved in increasing access to engineering education to all school leavers who are able including those coming from rural background or economically or socially disadvantaged families through policy initiatives on opening new institutions both by public and private sectors and through regulating fees, and grant of scholarships to economically weaker students.

— While access and equity issues can be handled through allocation of sizeable funds and building physical infrastructure where desirable, the quality issue would need major effort over at least the next three plan periods as this would involve competence development of existing teachers, attracting through suitable incentives a larger number of engineering graduates and post-graduates to a teaching and research career, synergizing the innovation efforts and potential of educational institutions, research laboratories and industry, and above all a change in the mindset of educational managers and policy planners in removing barricades to achieving excellence.

15.2 ETHICS OF ENGINEERING PRACTICES

— Knowledge of professional codes of ethics and importance of professional propriety should be integrated into the professional education system so that these are cultivated and nurtured by sensitization as a part of professional orientation of our students.

— The privilege of the profession demands unconditional adherence to defined codes of practice and to its moral and ethical values. All academic institutions must have in-house monitoring and regulatory mechanism for ethical values through suitable "Ethics Vigilance" or "Knowledge Integrity" committees. Further,

— There is a need for a national, non-government, and quasi-judicial committee/ commission to provide a think tank and a watchdog to deal with unethical cases of national importance.

— The professional responsibility to adhere to ethical values must be supported by legal, moral, social and financial rights and also due recognition by government , employers, professional societies, colleagues and society for the individual. A whistle blower's act needs to be adopted nationally for respecting and protecting the whistle blower who exposes misconduct or unethical practices in an organization.

15.3 FACULTY RELATED ISSUES

— In order to enhance the scope of learning, AICTE and other Government Institutions have to adopt a proactive role. The QIP is a very successful initiative. During the next phase, QIP has to take the responsibility of grooming a large number of faculty members across the country. It has to review and redefine goals and targets for continuous improvement of qualified teachers in adequate numbers. In order to encourage innovation in teaching, teaching methodology and research, the issue of career advancement of the teachers at various engineering colleges is of paramount importance. The Quality Improvement Programme (QIP) has to enter into a new paradigm. The QIP scholars train themselves well and pick up ideas during their stay in the Institutes of higher learning. Often they become stagnant on their return to the parent Institutes because of lack of facilities and encouragement. It is suggested that each QIP scholar, who has successfully completed PhD, be given a sponsored research project to execute. The colleges have to understand that faculty with only a Master's education will not be able to bring achievement on a par with the international level. The quality of technical education in the entire country will reach a level of excellence, if the morale of the teaching community is bolstered by enthusiasm, innovation and inquisitiveness. The society has to recognize the importance of life-long learning.

15.4 CURRICULA ISSUES

— Taking an overall view, engineering education in the country must have a minimum but essential common structure and features. This will be necessary to assess the level of competence, which the education system enables the graduating students to achieve for their mobility and for the recognition of their degrees and diplomas. The common undergraduate structure should prescribe the duration of the course of study, the minimum number of course credits and a flexible framework of distribution of the courses in terms of minimum requirements for humanities, sciences, core engineering subjects, electives for specialization and other requirements such as projects, self-study and value addition courses. These courses should also have coherence and reflect a unity of purpose. The current trend world over is to structure the academic programmes in a Credit based academic system. The idea of a Credit based system is that a student has to earn a prescribed minimum number of Credits in order to fulfill the requirements of graduation.

— The programmes that would be offering postgraduate degrees must have adequate faculty members and each one of them should be well qualified to teach postgraduate courses. The faculty members have to be active in research and development (R&D) activities. They should have some sponsored/ industrial research projects. Securing patents is highly encouraged. The R&D activities will be substantiated by writing papers and (or) books. The faculty members should be capable of developing and floating new postgraduate courses in the emerging areas. The country will need huge manpower during next ten years to sustain the academic growth. The focus will be to train at least some of the postgraduate students for a career in teaching. The postgraduate students should be encouraged to assist the teachers in undergraduate tutorials and laboratories. Having completed the programmes, the postgraduate students should be able to demonstrate teaching abilities.

— More electives courses need are to be introduced. The electives should cover almost all new emerging area of sciences and technology. This coupled with challenging problems for research will attract motivated students. The combined effect will offset the scarcity of master's students.

15.5 TEACHING LEARNING PROCESSES

— The first four semesters should cover mathematical, physical, chemical and life sciences. There should be coverage on environmental sciences. The "humanities courses" must carry close to 10% of the total credit requirement. The end-semester examination should not carry more that 50% weightage. The mid-semester examination, quizzes and assignments should carry 50% weightage. Emphasis should be there to follow continuous evaluation system. The degree should be awarded on a credit based approach. Besides compulsory "engineering science" courses, there should be some optional courses in "engineering sciences" in order to provide overall exposure to various branches of engineering. During the fifth semester, a student will be encouraged to select one area in the other streams (Minor) and if he/ she earns credits for three courses (including departmental and open electives) in that area, the student will graduate in the "Major" together with a chosen "Minor". The students should undergo courses designed to improve "skills" such as software skills, accounting, hardware skills and instrumentation. There should be an effort to blend the subjects, such as biotechnology, bioinformatics, genetics etc. with the mainstream subjects of various streams. There should be significant weightage for the laboratory courses and field work. The laboratory courses should have some challenging experiments where a student would be required to design and fabricate the experiment. The laboratory facilities should be motivating for learning and justify choosing engineering as a profession. Emphasis has to be there to develop all-round communication skills. Students should be taught to nurture a wide range of communication tasks, staring from presenting experimental data to presenting any theme or concept verbally.

15.6 GOVERNANCE AND MANAGEMENT

— An autonomous Higher Education Council/ Commission, consisting of prominent stake holders, is needed to act as a national think-tank for planning a world class higher and technical education in the country. Further,

— Roles of UGC and AICTE need to be redefined and confined to development, standardization, monitoring and regulation.

— Credible accreditation of higher education as per international standards should be handled by one or more non-government, financially autonomous, self-sustaining and transparent agencies drawn from professional and industrial organizations.

— India needs a large number of world class institutions which would require participation and collaboration of public, private, and foreign players for which a suitable legislation and level playing field need to be in place.

— A National Education Development Bank should be set up to provide loans to students at nominal rates of interest, with judicial empowerment to recover loans.

— The composition and the functions of the governing bodies of technical institutions, both public and private, need to be reviewed. Membership should be drawn from a database of well known educationists, professionals, industrialists and alumni who have a demonstrated stake in the system. Monitoring and grading of the performance of an institute should be the major responsibility of the governance system.

— Senate or an equivalent apex body for all academic matters should be small and compact for effective planning, monitoring, and decision making

— Quality of faculty determines to a large extent the quality of the institution and its product, namely students. Any type of implicit or explicit reservation would be detrimental to academic excellence. Affirmative action policy to ensure national homogeneity and gender equality must, however, be pursued.

— Performance incentives should be provided to faculty in the form of additional increments, perks, or honorarium from their sponsored projects, consultancy, or industrial work.

— Heads of Institutions should be invited to the position on the basis of recommendations of search committees consisting of distinguished academicians and engineering professionals.

— Research-oriented academic institutions should be treated as "not-for-profit" trusts which must plough back profits into the institute.

— An autonomous and self-sustaining National Testing Service for both UG and PG admissions on a continuing basis needs to be nurtured by premier institutions to replace the existing multitude of such systems.

— All technical institutions should be incentivised to follow a similar system of continuous evaluation, credits and semesters, with in-principle possibility of inter-institutional migration and mobility of students.

— Liberal policies are needed to encourage foreign students on our academic campuses which would challenge our teaching -learning process, change the cultural ecology, and earn us a brand name.

15.7 RESEARCH, DEVELOPMENT AND INDUSTRY INTERACTION

— If India wants its best graduates to do postgraduate studies and research in its premier institutions, it has to compete with the West in providing attractive fellowships, excellent research facilities, stimulating environment, competitive and challenging research focus, and opportunity to work and to be absorbed by the knowledge institutions or industry . Further,

— The problems associated with R&D projects sponsored by funding agencies are many which include lack of national databank on projects and their outcome, lack of competitive bidding on projects, lack of transparency, assessment and accountability, and notoriously erratic financial rules and bureaucracy, etc. These enigmas must be resolved if research has to become an attractive, effective and satisfying experience for any engineering faculty. A paradigm shift to a transparent, flexible, liberal, pragmatic, accountable and autonomous system is urgently needed.

— Major attitudinal and mindset changes such as mutual trust, respect, motivation, spirit of cooperation etc need to be inculcated among all the stakeholders in industry and the academia.

— Industry-Academia collaborating partners should devise liberal and flexible rules, and fast-decision making process for R&D collaborations, recruitment, exchange visits, lectures, conferences, awards, incentives and rewards. A mechanism for involvement of partners in the management and assessment of project and its outcome in a business-like manner needs to be evolved.

15.8 TECHNOLOGY ASSISTED LEARNING

— The e-learning framework must provide access to knowledge and related data anytime anywhere. The e-learning environment must provide a flexible workflow and process model that can be fine-tuned and configured to meet the needs of the organization. The e-learning framework must allow additional components to be integrated easily using open software. The e-learning framework must allow content and other data to be exchanged and shared by tools and systems connected via internet. Through the course management tools (Learning Management Systems), syllabi and other education information are to be made more accessible to the students.

15.9 INNOVATION, INTELLECTUAL PROPERTY RIGHTS, AND ENTREPRENEURSHIP

— Although institutions of higher learning are ideal propagators and promoters of enquiry, research and critical evaluation, their success in producing innovators depends to a large extent on the attitude and mindsets of their students. Innovation and creativity have to be fostered in the student's attitude and learning methodology right from their childhood. The role of the schoolteacher both at the primary and secondary levels is the key to developing among the students an enquiring attitude and problem-solving ability. Rote learning and accepting ideas without questioning at the school level can become a stumbling block for real creative and innovative work at the higher education level since students with such a background would not feel encouraged to traverse forbidden paths and look at newer ways of tackling problems. Roadblocks to success can come also if faculty in the colleges are not fully committed to imbibing new knowledge and advancing frontiers of knowledge, and the college management does not recognize and reward creative and innovative achievements. Effectiveness of university innovation and promotion policy depends largely on free access to needed facilities- be it published literature, laboratory/ field experimentation and peer interaction, persistence and ability to learn from failures, availability of critical but fair assessment of effort, assurance for protection of intellectual property rights, and a proper recognition and reward system.

— Engineering Education offers enormous opportunities for promoting creativity and innovation since the processes of innovation are identical to those of problem solving and design which are the basic foci of all engineering education. Fostering a culture of innovation among students would require training in critical thinking, encouraging asking inconvenient questions in the classroom, thinking out of the box, looking at problems from multiple points of view, generating ideas and solutions including those which appear at first sight to be highly improbable, providing access to experimentation and other institutional facilities after normal working hours, frequent brainstorming among themselves with or without guidance of a teacher.

— AICTE should develop and establish a consortium of Engineering institutions devoted solely to encouraging, fostering and supporting innovative ideas generated by students and faculty and converting them to commercial products with the support of interested industry or through venture funds.

— It is recommended that all engineering institutions have an IPR promotion cell which will give regular training to students, research scholars and faculty in protecting intellectual property and equip them with tools to evaluate their research outputs for eligibility for patenting and commercial exploitation.

—   Academic institutions must bring about attitudinal and mindset changes among all the stakeholders, namely, faculty, students, alumni and support staff to accept entrepreneurship as an important academic and R&D activity of the institution. Faculty members and students should be encouraged to participate in entrepreneurial activities as advisors, consultants, or part-time partners as long as their primary academic responsibilities are not compromised. To be viewed as an investment in the economic future of the country, the government should fund entrepreneurial activities liberally through grants, fellowships, and awards. The institutions should adopt simple, liberal and pragmatic and dynamically flexible rules of business/governance for entrepreneurial activities.

—   All institutions should be encouraged to set up STEPs (Science and Technology Entrepreneurship Park), EDCs (Entrepreneurship Development Cells), or TBIUs (Technology Business Incubators) and to allow faculty to mentor entrepreneurs and join the ventures so created as stakeholders.

—   All institutions should be allowed to establish a corpus or venture capital out of the consultancy income or donations to invest in the incubated businesses on equity basis or joint ventures. The academic institutions should be given the status of not-for-profit Societies so that any profit is ploughed back into the institution in whatever form it deems fit.

ANNEXURE-1

Based on Annexure referred to in part (a) of the reply to the Lok Sabha Unstarred Question No.56 for 17.2.2009 asked by Shri Jivabhai Ambalal Patel, Shri V.K.Thummar & Shri Subhash Maharia, M.P.s regarding Setting up of Technical Institutions.Number of Degree Level Technical Institutions Approved by AICTE as on 31/08/2008

S.No.	States/UTs	Engg & Tech.	MBA	MCA	Pharmacy	HMCT	Applied Art & Crafts	PGDM
Central								
1.	Madhya Pradesh	161	56	47	93	4	0	7
2.	Chhatisgarh	41	7	7	11	0	0	2
3.	Gujarat	55	51	26	75	1	0	11
	Total Central	**257**	**114**	**80**	**179**	**5**	**0**	**635**
East								
4.	Mizoram	1	0	1	1	0	0	0
5.	Sikkim	1	1	0	1	0	0	0
6.	Orissa	68	29	37	15	2	0	15
7.	West Bengal	71	27	27	10	4	0	2
8.	Tripura	3	0	1	1	0	0	0
9.	Meghalaya	1	0	1	0	0	0	0
10.	Arunchal Pradesh	1	0	0	0	0	0	0
11.	Andaman & Nicobar	0	0	0	0	0	0	0
12.	Assam	7	5	3	2	0	0	1
13.	Manipur	1	1	0	0	0	0	0
14.	Nagaland	1	0	0	0	0	0	0
15.	Jharkhand	13	4	2	1	0	0	3
	Total East	**168**						

North								
16.	Bihar	15	11	6	4	0	0	1
17	Uttar Pradesh	241	125	101	91	10	1	88
18.	Uttaranchal	19	23	13	14	7	0	2
	Total North	**275**						
North West								
19.	Chandigarh	5	0	2	1	0	1	1
20.	Haryana	116	56	29	34	3	0	10
21.	Jammu & Kashmir	7	9	3	1	0	0	0
22.	New Delhi	19	13	18	6	1	1	24
23.	Punjab	70	55	24	38	8	0	4
24.	Rajasthan	81	49	19	54	8	0	15
25.	Himachal Pradesh	9	8	1	11	1	0	0
	Total North West	**307**						
South								
26.	Andhra Pradesh	527	231	385	258	2	0	24
27.	Pondicherry	9	1	6	1	0	1	0
28.	Tamil Nadu	352	154	208	43	1	0	4
	Total South	**888**						
South West								
29.	Karnataka	157	109	73	80	20	0	15
30.	Kerala	94	37	38	33	4	0	7
	Total South West	**251**						
West								
31.	Maharashtra	239	168	58	120	10	6	48
32.	Goa	3	1	1	2	0	1	1
33.	Daman & Diu, Dadar, NH0		0	0	0	0	0	0
	Total West	**242**						
	Grand TOTAL	**2388**	**1231**	**1137**	**1001**	**86**	**11**	**285**

ANNEXURE-2

Based on Annexure referred to in part (b) of the reply to the Lok Sabha Unstarred Question No.56 for 17.2.2009 asked by Shri Jivabhai Ambalal Patel, Shri V.K. Thummar & Shri Subhash Maharia, M.P.s regarding Setting up of Technical Institutions.Number of proposals received for Establsihment of New Technical Institutes for the Academic year 2009-2010 upto 31.12.2008

Region	State	Engg.	PGDM	MBA	MCA	Pharmacy	HMCT	Applied Arts	Total
Central	Madhya Pradesh	50	16	63	7	3	0	0	139
	Chhatissgarh	10	2	7	1	4	0	0	24
	Gujarat	43	8	67	18	14	1	0	151
East	Orissa	53	10	24	6	1	0	0	94
	Assam	9	0	5	0	0	0	0	14
	Meghalaya	0	0	0	0	0	0	0	0
	Manipur	0	0	0	1	0	0	0	1
	Mizoram	0	0	0	0	0	0	0	0
	West Bengal	23	3	7	1	1	0	0	35
	Nagaland	0	0	0	0	0	0	0	0
	Jharkhand	2	3	1	0	1	0	0	7
	Sikkim	1	0	0	1	0	0	0	2
	Arunachal Pradesh	0	0	0	0	0	0	0	0
	Tripura	1	0	0	0	0	0	0	1
	Andeman & Nico	1	0	0	0	0	0	0	1

Region	State								
North	Uttar Pradesh	83	84	130	11	11	4	1	324
	Uttranchal	13	3	14	3	1	1	0	35
	Bihar	12	2	3	2	0	0	0	19
North West	Chandigarh	0	0	1	3	0	0	0	4
	Delhi	1	3	2	0	1	0	0	7
	Haryana	38	11	36	9	1	3	0	98
	Himachal Pradesh	11	0	6	4	0	1	0	22
	J&K	0	0	0	1	0	0	0	1
	Punjab	16	1	30	10	3	1	0	61
	Rajasthan	49	12	65	4	7	6	0	143
South	Tamil Nadu	144	3	38	7	3	0	0	195
	Pondicherry	4	0	0	0	0	0	0	4
	Andhra Pradesh	176	31	178	9	49	2	0	445
South West	Karnataka	32	18	26	4	0	0	0	80
	Kerala	29	1	8	1	2	0	0	41
West	Maharashtra	85	37	123	17	22	3	0	287
	Goa	0	2	0	0	0	0	0	2
	Daman & Diu	0	0	0		0	0	0	0
	Grand Total	**886**	**250**	**834**	**120**	**124**	**22**	**1**	**2237**

ANNEXURE-3
Statement showing NIT - wise details of intake during 2007-08

Sl. No.	Name of the Institute	Under Graduate Level		Post Graduate Level	
		Annual Intake	Total Students Strength	Annual Intake	Total Students Strength
1.	NIT, Agartala	266	1013	—	—
2.	MNNIT, Allahabad	530	1924	372	314
3.	MANIT, Bhopal	450	2029	328	336
4.	NIT, Calicut	613	1892	324	669
5.	NIT, Durgapur	520	1936	164	122
6.	NIT, Hamirpur	330	1158	162	102
7.	MNIT, Jaipur	520	1926	198	353
8.	NIT, Jalandhar	490	1524	210	206
9.	NIT, Jamshedpur	390	1600	58	200
10.	NIT, Kurukshetra	540	1672	298	445
11.	VNIT, Nagpur	480	1768	239	252
12.	NIT, Patna	400	1104	126	83
13.	NIT, Raipur	620	2480	129	318
14.	NIT, Rourkela	420	1463	342	491
15.	NIT, Sikhar	300	972	90	57
16.	NIT, Srinagar	410	1054	64	77
17.	SVNIT, Surat	450	1719	301	295
18.	NITK, Surathkal	480	2045	293	421
19.	NIT, Tiruchirapalli	608	2367	438	919
20.	NIT, Warangal	480	1571	433	745
	Total	9297	33217	4569	6405

ANNEXURE-4

Details of Faculty position, Student Strength, Student Intake and Number of Ph.D.s awarded in IITs during the Academic Session 2007-08

Indian Institutes of Technology (IITs)	Kharagpur	Bombay	Madras	Kanpur	Delhi	Guwahati	Roorkee	Total
	1	2	3	4	5	6	7	8
Year of Admission of 1st batch	1951	1958	1959	1960	1961	1995	2001*	
Faculty strength	519	433	392	311	419	193	367	**2634**
Student strength	6625	5420	5011	3810	4995	2126	4402	**32389**
Student intake (UG/PG) in 2007	1901	1550	967	914	1550	796	1467	**9145**
Number of PhDs Awarded in 2007	167	152	125	86	145	16	107	**798**
No. of patents applied for.	13	10	13	24	14	5	NIL	**79**

***year of conversion into an IIT.**

About the Authors

Prof. Gautam Biswas is currently on deputation as the Director of Central Mechanical Engineering Research Institute (CMERI), Durgapur. Before joining CMERI, Prof. Biswas was holding the post of GD and VM Mehta Endowed Chair Professor of Mechanical Engineering at the Indian Institute of Technology Kanpur. He has earlier been the Dean of Academic Affairs at IIT Kanpur. He was a Humboldt Fellow in Germany and JSPS invited fellow in Japan. He is a Fellow of the American Society of Mechanical Engineers (ASME), the Indian National Academy of Engineering and the National Academy of Sciences (India). He is also a Fellow of the Institution of Engineers (India). He works in the area of Computational Fluid Dynamics and Heat Transfer. He has developed methodologies for handling free surface flows, bubble formation in film boiling and turbulent transport.

Prof. K. L. Chopra is a former Director of Indian Institute of Technology Kharagpur. He is a former Professor of Physics, Head, Thin Film Solid State Technology Laboratory, Physics Department and Centre for Energy Studies, and Dean, PG Studies and IRD at Indian Institute of Technology Delhi. He is the author/ co-author of several scientific books and is a recipient of such awards as Max Planck Fellowship, S.S. Bhatnagar Prize, Bhabha Award, Mahalanobis Medal, Aryabhatta Medal, etc. Prof. Chopra is a Fellow of the Indian National Science Academy, Indian Academy of Sciences, National Academy of Sciences (India) and the Indian National Academy of Engineering. His research areas include: Solid State Physics and Technology of Thin Films (now known as Nanotechnology), Surface Science, Thin Film Solar Cells, High Temperature Superconductors and Amorphous Materials. He received citation laureate award from ISI, USA for being "one of the seven most highly cited Indian scientists".

Prof. C.S. Jha was a former Director of Indian Institute of Technology Kharagpur and former Vice Chancellor of Banaras Hindu University. Before taking over the Directorship of IIT Kharagpur, Prof. Jha was a Professor of Electrical Engineering at the Indian Institute of Technology Delhi. He was also Education Adviser (Tech) of Government of India. Prof. Jha was a Fellow of the Indian National Academy of Engineering and the Institution of Engineers (India). He did extensive work in the areas of electrical machines and drives, control systems and energy conversion. A well known academician, administrator and dedicated educator, Prof. C.S. Jha breathed his last on June 16, 2009.

Prof. D.V. Singh is a former Director of IIT Roorkee and former Vice Chancellor of University of Roorkee. Earlier he was the Vice Chairman of AICTE and also the Director of Central Road Research Institute (CRRI), New Delhi. Prior to joining CRRI, he was a Professor of Mechanical Engineering at the University of Roorkee. Prof. Singh is recipient of S.S. Bhatnagar Prize, Khosla Research Prize, INAE Jai Krishna Memorial Award and Distinguished Alumnus Award of IIT Roorkee. He delivered INSA G.P. Chatterjee Memorial Lecture. His research includes dynamic stability of vehicles and tyre pavement interactions. He has done extensive work on fluid film seals and lubrication to include hydrostatic and hydrodynamic bearings with complexities of fluid rheology, presence of contaminants, a wide range of variations in bearing and seal geometry, micro-polar fluids, piezothermoviscous effects, gas bearings in laminar and super-laminar regime and magnetohydro-dynamic bearings. He had excursions in Physics of welding process, arc stability and effect of welding parameters on the microstructure of weldments and heat effected zones. Prof. Singh is a Fellow of the Indian National Academy of Engineering, Indian National Science Academy, Indian Academy of Sciences and the National Academy of Sciences (India).